Renovation Problems in Constructions and Historic Buildings

Renovation Problems in Constructions and Historic Buildings

Editor

Krzysztof Schabowicz

MDPI

Basel • Beijing • Wuhan • Barcelona • Belgrade • Novi Sad • Cluj • Manchester

Editor
Krzysztof Schabowicz
Faculty of Civil Engineering
Wroclaw University of
Science and Technology
Wrocław
Poland

Editorial Office
MDPI AG
Grosspeteranlage 5
4052 Basel, Switzerland

This is a reprint of articles from the Special Issue published online in the open access journal *Materials* (ISSN 1996-1944) (available at: www.mdpi.com/journal/materials/special_issues/Renovation_Historic_Buildings).

For citation purposes, cite each article independently as indicated on the article page online and as indicated below:

Lastname, A.A.; Lastname, B.B. Article Title. *Journal Name* **Year**, *Volume Number*, Page Range.

ISBN 978-3-7258-1634-7 (Hbk)
ISBN 978-3-7258-1633-0 (PDF)
doi.org/10.3390/books978-3-7258-1633-0

Contents

About the Editor

Krzysztof Schabowicz

Krzysztof Schabowicz is a specialist in the fields of building engineering, building law, diagnostics, and maintenance of engineering structures. He deals with ventilated façades, in particular production technology and testing of external fiber cement cladding in the field of detection, identification, and classification of degradation and damage processes, as well as the methodology of these tests. He conducts scientific research and development works related to the implementation of non-destructive diagnostic devices and technologies in construction works, including the use of artificial intelligence.

Author and co-author of 5 books, over 300 publications, and 10 patents. He has more than 1000 citations in the Web of Science. He serves as an Editor of *Materials* (MDPI) and Editorial Board member of *Civil Engineering and Architecture* (HRPUB) and *Nondestructive Testing and Diagnostics* (SIMP). He developed more than 500 reviews of journal and conference articles. He is the co-author of nine patents and one patent application. He is a member of the Polish Association of Civil Engineers and Technicians (PZITB) and a member of the Polish Association of Building Mycology (PSMB). His research interests include concrete, fiber-cement, ultrasonic tomography, impact-echo, impulse-response, GPR and other non-destructive tests, and artificial intelligence.

In 1996, he obtained an M.Sc. (Hons.) in Civil Engineering from Wrocław University of Science Technology (WUST) in Poland, a Ph.D. (Hons.) in Civil Engineering from WUST in 2003, a D.Sc. in Civil Engineering from WUST in 2015, and a title of professor in 2020. Currently, he works as a professor at the Faculty of Civil Engineering of the Wrocław University of Technology in the Department of General Building Engineering.

Preface

The field of renovation problems in construction and historic buildings is very broad and has given rise to several fascinating engineering and scientific perspectives.

This Special Issue was proposed and organized with the intention to present recent developments in the field of renovation problems in constructions and historic buildings.

The articles highlighted in this Special Issue relate to different aspects of the renovation problems in construction and historic buildings.

It was my pleasure to invite manuscripts for this Special Issue, which mainly focused on novel different renovation problems in constructions and historic buildings.

Krzysztof Schabowicz
Editor

materials

MDPI

Article

Material Testing of Historic Bricks and Mortars in Degraded Masonry Structures

Dariusz Bajno [1], Krzysztof Schabowicz [1] and Agnieszka Grzybowska [2,*]

1 Department of Building Engineering, Faculty of Civil Engineering, Wroclaw University of Science and Technology, 50-370 Wrocław, Poland; dariusz.bajno@pwr.edu.pl (D.B.); krzysztof.schabowicz@pwr.edu.pl (K.S.)

2 Department of Building Structure, Faculty of Civil and Environmental Engineering and Architecture, Bydgoszcz University of Science and Technology, 85-796 Bydgoszcz, Poland

* Correspondence: agnieszka.grzybowska@pbs.edu.pl

Abstract: The subject of this article is material research carried out on the ruins of a medieval castle located in west-central Poland. This facility was built at the beginning of the 15th century by the Order of St. John, and during its long life, it was subjected to many reconstructions. Unfortunately, in 1975, it was destroyed by fire. Since then, it has been left in a state of advanced ruin, exposed to climatic influences without any protection. The subject of the research was to assess the possibility of maintaining such buildings in a severely degraded condition while ensuring their technical efficiency. The article discusses a particular instance of "consolidation" applied to a structure in a state of historical, architectural, and structural ruin. After the diagnosis, it was determined that the structure should be safeguarded using a minimally invasive method. The purpose of these activities was to answer the question of whether the structure could be left to continue operating despite failing to meet the requirements of current standards and regulations while posing an additional danger to itself and the environment,. This goal was achieved by obtaining a considerable amount of data on the condition of the materials embedded in the masonry structure, thanks to which the initial parameters for conducting an assessment of the technical condition of the damaged masonry structure and evaluating the degree of its danger were developed. The results of the research and analysis carried out and described in this article can be used in other similar situations where saving national heritage objects through "artificial modern" strengthening will be unsafe and will lead to a loss of their authenticity. We still have a long way to go to develop a comprehensive method for "in situ" diagnosis of heterogeneous masonry structures, so we should use possible techniques and knowledge to conduct such assessments and propose rescue methods for historically valuable objects in a way that could minimize the damage and that can "easily" disappear from our surroundings. Each study should have a specific purpose, not only research but also a long-term perspective, making it possible to leave material for further research and analysis, including testing new research methods in real conditions of its installation.

Keywords: permanent ruin; nondestructive testing; mycological research; scanning; masonry structures; structural and material protection

Citation: Bajno, D.; Schabowicz, K.; Grzybowska, A. Material Testing of Historic Bricks and Mortars in Degraded Masonry Structures. *Materials* **2024**, *17*, 3192. https://doi.org/10.3390/ma17133192

Academic Editors: Antonio Caggiano and Sukhoon Pyo

Received: 6 May 2024
Revised: 19 June 2024
Accepted: 20 June 2024
Published: 29 June 2024

1. Introduction

Building ceramics is one of the oldest construction products [1,2]. Ordinary ceramic bricks were produced in Egypt before 4000 BC, while colored ceramic bricks were produced around 3000 BC. In Europe and Poland, the first stone and brick structures were built from hand-formed bricks (the so-called finger bricks) already in the 10th century. The mechanized production process began in the 18th century, while fully mechanized industrial production involving the use of a ring furnace and a mechanical press took place in the second half of the 19th century. The brick firing process takes place at temperatures of

800 to approximately 1200 °C and over 1500 °C in the case of refractory ceramics. Signs informing about their producers have been embossed on bricks since ancient times, and this principle is also used today. The basic building materials of most historical buildings were wood, stone, ceramics, and lime. While the technical condition of, e.g., wooden structures can be assessed quite accurately on the basis of dedicated acoustic and resistographic methods (measuring cutting resistance) among others, the same research techniques will not necessarily work in the assessment of masonry structures. Research on diagnostic methods for walls, including historical ones, is still at the stage of tests and analyses both in situ and in laboratories. So far, no effective method for assessing these structures has been developed [3–11]. The relationship between the history of the object and, especially, its last stage, when it was subjected to loads not expected of it, is close because it is the basis for the genesis of its technical wear and tear. This is not just about laboratory tests, for which it is possible to collect and destroy any number of samples. In the case under study, we are dealing with material built into a large-scale facility. Here, the results of possible tests as well as our knowledge and well-made decisions will determine the safety of the facility left for further operation.

2. Case Study

Each building structure is subject to aging processes, i.e., it wears out technically, environmentally, and functionally [1,12–14]. The pace of such processes depends on the awareness and care of these structures, as well as on the technology used to make them, external influences, and random events. Mainly, attention to the condition of buildings and structures will determine their technical and operational life [15]. The process of technical wear and tear of each building begins at the commencement of construction works and will always have a rapidly progressing tendency unless it is slowed down by appropriate handling, including systematically diagnosing the technical condition and allowing only justified changes in its structure, supported by appropriate analyses and computational simulations [1,12,13,16,17].

The inspiration to take up the topic presented in the article is very frequent abandonment of fragments of historic fortifications in a state of advanced degradation and, thus, exposing them to further destructive influence of the environment [14]. In such situations, it becomes necessary to "prepare" them for further use in the above-mentioned environment while ensuring the required level of reliability [18–20]. One example of such structures is the remains of a medieval castle (Figures 1–3). In the past, it was the seat of the commander of the Order of St. John. It was built in the years of 1545–1564, replacing the wooden knight's manor that existed there in the years of 1426–1429. The castle in Słońsk has been expanded and rebuilt many times since its construction. In 1652, it was burned down by the Swedes, and its last expansion took place in 1783 (Figure 1a). Due to a devastating fire that took place in 1975, the facility became a ruin left for over 45 years, threatening a construction disaster. Because it is a monument with high tourist potential, closely related to the rich history of Brandenburg, the Netherlands and Europe, it was decided to preserve it for future generations in its current state (Figure 1b,c).

(a) (b) (c)

Figure 1. View of the castle: (**a**) in 1896 [21], (**b**) in 2018 [22], and (**c**) in 2023 (photo by authors).

Figure 2. View of walls and vaults—as of 2018 (photos by authors).

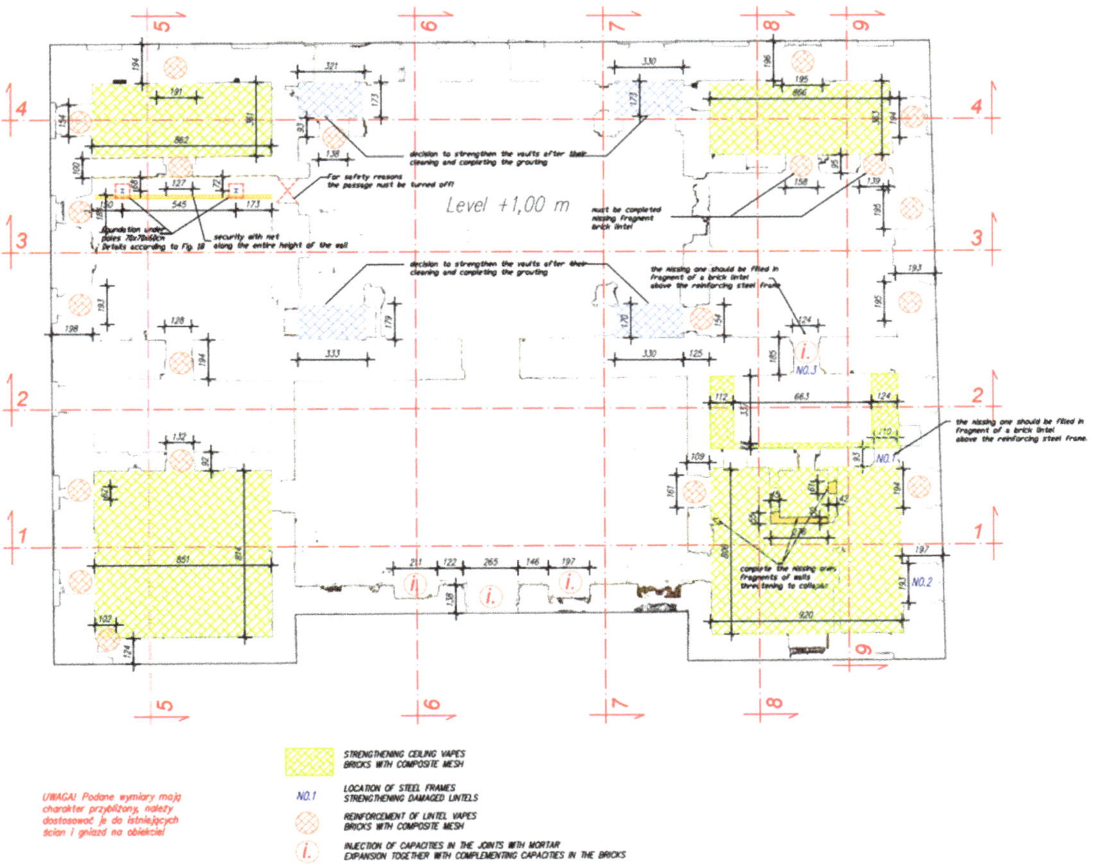

Figure 3. Three-dimensional scanning of the building at a height of +1.0 m above the ground floor (own study and [23]).

The medieval walls of the original structure of the facility have survived to this day, and it was decided to secure them. Already in 2018, the technical condition of the castle's masonry structure was considered disastrous. The threat here was the exposed slender walls 19 m high without stiffening in the form of vaults, wooden ceilings, and a roof. Chimney walls were particularly at risk of losing stability, while the remains of the vaults contained a large layer of rubble and vegetation growing on them (Figure 2). In 2021, the building was protected with a temporary wooden roof and covered with roofing felt, which resulted in the quick drying of bricks and mortars with intensive ventilation caused by the lack of windows and doors. The remains of the vaults, like the rest of the building, were on the verge of a construction disaster. The above-mentioned loss of wooden ceilings, vaults, and large fragments of walls significantly weakened the stiffness of the remaining structure of the building, which was due to the walls being left unprotected for the last 45 years. The facility required urgent protection and reinforcement of walls and lintels, as well as reinforcements and additions to the vaults. Carrying out this work turned out to be an extremely difficult and dangerous activity due to the additional threat associated with the not fully recognized and heavily debris-covered surfaces of the structure, which already lost its character as a building. Debris was lying on the ground floor and on the remains of the vaults, which made it very difficult to inventory places that were then inaccessible, hence scanning the facility [23] turned out to be very helpful as it did not have archival documentation (Figure 3).

By the decision of the monument protection services, the facility was to remain in ruins and intended for sightseeing, without supplementing or rebuilding the deformed structures. The owner of the castle obtained consent only to protect the damaged structures against construction disasters and eliminate the threat to visitors, which turned out to be difficult to implement due to the fact that the standard conditions [18] would not be met both in relation to the requirements for modern building materials and the stability conditions. As mentioned above, a helpful solution was to scan the entire facility to determine the extent of deformation of the remains of its walls and vaults. The first very difficult stage was the removal of rubble from both the ground floor of the castle and its first aboveground floor where the vaults and their fragments remained due to the threat posed by the deformed and heavily strained masonry structure. The benefit of carrying out the above-mentioned activities included not only cleaning up the rubble but also recovering approximately 70% of valuable historical bricks, which were used to supplement and strengthen the structures of walls and vaults. Defective and weakened historic ceramic bricks with a tendency to delaminate, peel, and crumble were replaced with other bricks coming from elements that no longer existed here, and the bricks themselves were preserved in good condition, placed on a sand-lime mortar.

In Figure 3, the green color indicates the vaults that remained in the structure of the building in whole or in part, while in Figure 4, the deformed chimney wall that tilted the most from the vertical is marked.

Testing bricks or other materials cannot be an end in itself when it comes to an existing and still used facility.

The brick itself is only part of the knowledge of the technical value of the masonry structure it creates. Low parameters of individual bricks may disqualify such a structure as a whole or have no major impact on it. Only the assessment of walls in facilities exposed to the negative impact of the external environment, which is a highly unfavorable interference for them, allows for a comprehensive assessment of the tested material, hence the research and analyses should assess the degree of wear of masonry elements and the possible threats this wear may pose. The heterogeneity of the bricks may disqualify the wall as a structure, but the distribution of stresses in the walls at an angle of 60° with the simultaneous elimination of concentrated loads and their considerable thickness (60–200 cm), as well as a dense network of vertical stiffeners, may also allow for further exploitation of the ruin after the introduction of an additional external structure (ensuring the stability of the walls and can be dismantled at any time) and, at the same time, not reducing the authenticity of the monument. Verification calculations for walls with a height of 19 m showed that the compressive stresses in the wall should not exceed the permissible values, even for bricks and (mortar) with a strength of 5 MPa and (1 MPa), and may even be lower by 15%. For 15 MPa class bricks, this reserve would be approximately 60%.

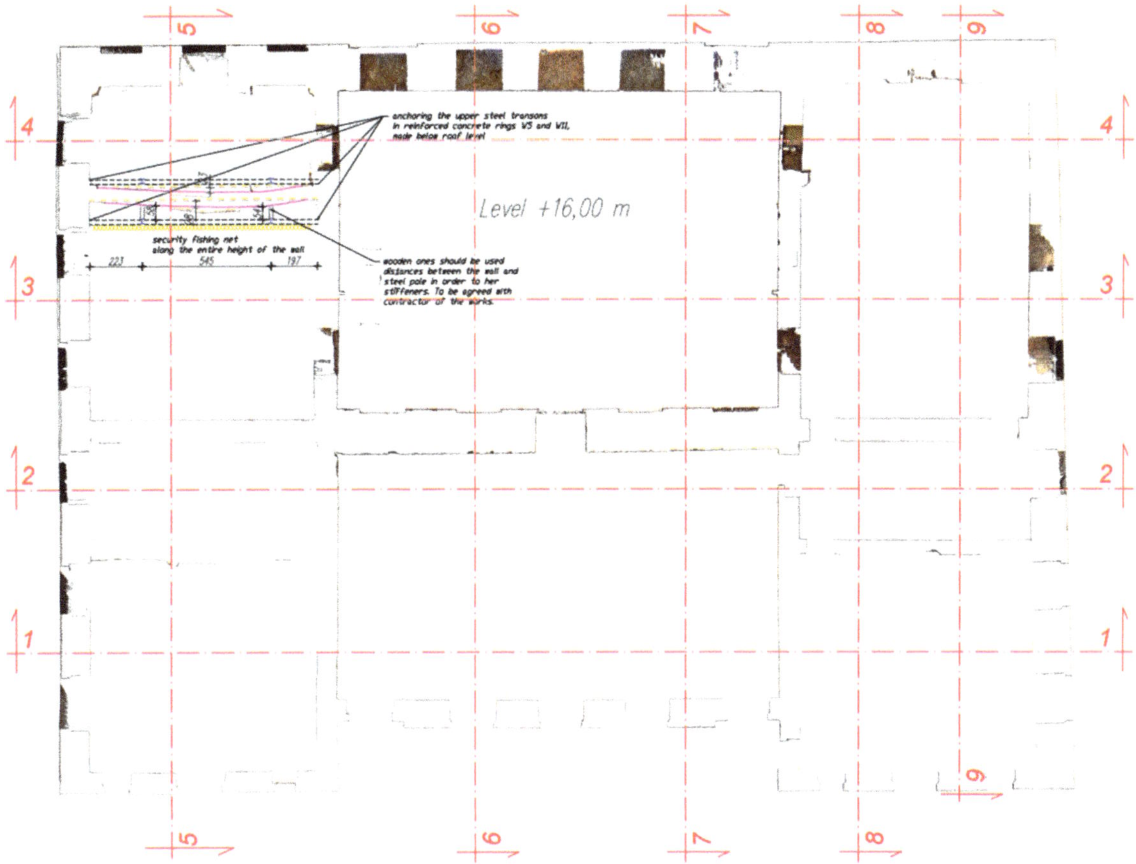

Figure 4. Three-dimensional scanning of the building at a height of +16.0 m above the ground floor (own study and [23]).

The conservation covered the vaults and walls but only on the surfaces where the bricks lost their structure. In order to strengthen, i.e., harden, the bricks and mortars in the masonry elements, it was proposed to introduce the silicate preparation Primer Hydro SF/Silikatfestiger into their structure, (art. no. 1072, Remmers) which is a liquid, mineral primer with a strong bonding effect. It is a colorless preparation that creates a structure permeable to water vapor, which structurally strengthens from 4 to 8 N/mm² by filling pores and small cracks with silica with a pH of approximately 11.5. It was applied by spraying three times until the substrate was fully saturated with it.

3. Review of the Literature

In the second half of the 20th century, the Schmidt sclerometer was first utilized to test building ceramics. This tool is still being tested to establish a close relationship between the number of rebounds (an assessment of impact energy) and the compressive strength of bricks and mortars. Although Ernst Schmidt, who patented the device in 1948, originally designed it to assess concrete structures within the measurement range of 10 to 70 N/mm², its application has expanded over time. Currently, work on the application of the above-mentioned instrument for strength testing of materials other than concrete is still based on its original purpose dedicated to concrete. An assessment of the effectiveness of these studies and analyses can be found in refs. [3–9]. They also constitute a theoretical and practical basis for "transferring" these techniques to other types of materials. Visual inspection should be an integral and basic element of the diagnostics of construction materials and products, as well as entire structures, which will initiate their further treatment. As mentioned above, there are a number of nondestructive methods for diagnosing historical buildings, dedicated mainly to materials used today, i.e., concrete, steel, and wood, which will not always be widely available or fully useful in specific situations. In Figure 5, the authors showcase the aforementioned continuously improved methods. These techniques are designed not only to assess the strength characteristics of elements and structures but also to detect hidden defects, especially in hard-to-reach areas. These defects can significantly impact the durability and safety of both the structures and their surroundings [4,6–8,10,11,24]. Knowledge of the location and size of such defects will allow for creating a picture of the damage and, thus, influence the accuracy of subsequent decisions and proceedings.

Already in the 1970s, Prof. Leonard Runkiewicz wrote about the imperfections of sclerometric tests of masonry structures. In his publication [25], Runkiewicz stated that his own research and the analysis of works [26–29] only indicated certain usefulness of N- and L-type Schmidt hammers for assessing the current strength of built-in bricks. He proposed his own, quite strict correlation between the strength and the number of rebounds of the N-type Schmidt sclerometer, expressed by Formula (1):

$$R_c = 0.305L^2 - 11.42L + 131.6 \text{ MPa}, \quad (1)$$

where L—number of rebounds.

In work [25], empirical dependencies for ceramic bricks were determined by Formula (2):

$$R_c = 0.203{L_s}^2 - 13L_s + 212.7 \text{ MPa}, \quad (2)$$

where L_s—reduced number of rebounds depending on the tension in the wall.

$$L_s = mL$$

where m = 1.00–0.75—coefficient at stresses from 0.0 to 2.0 MPa.

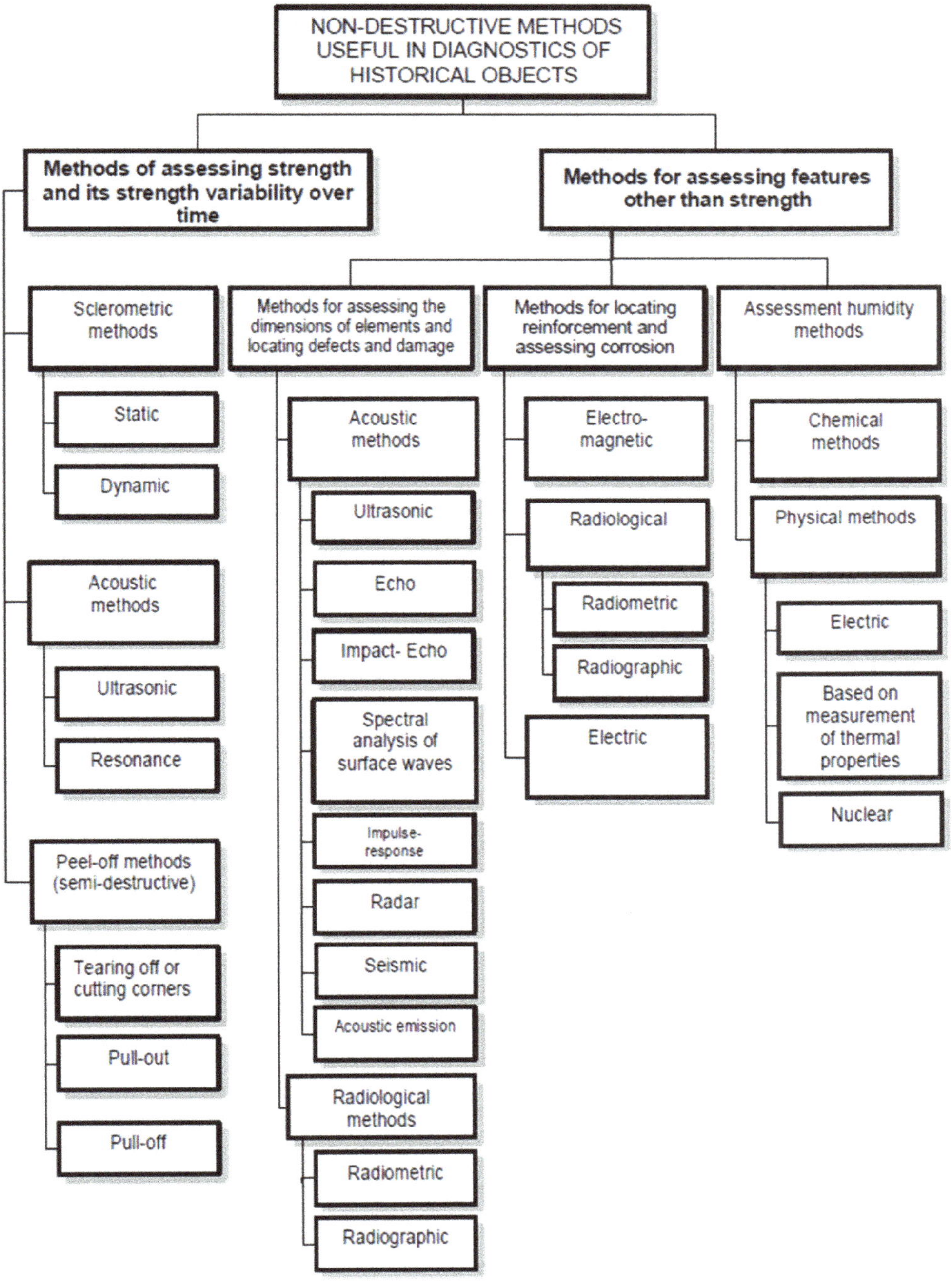

Figure 5. Nondestructive methods useful in diagnostics of historical structures [4–8,10].

The authors of the above article [25] stated in the summary that Schmidt hammers in testing historic buildings made of bricks could only be used to estimate strength [30]. When testing demolition bricks, due to the influence of small cracks and microcracks, the rebound numbers were reduced by approximately 20–30% compared with new bricks. Tests carried out on bricks by authors of [25] in historic buildings confirmed that small scratches and microcracks have a significant impact on the rebound numbers depending on their strength. It was found that there is a weak relationship between the surface hardness of the old brick and its compressive strength [25].

Nondestructive testing of building ceramics is still a topic of interest for researchers and engineers dealing with historical and contemporary structures, who strive to improve these methods.

The topic of nondestructive and low-destructive testing was discussed by Dawid Łątka in [31,32]. The author developed a unique correlation curve for determining the compressive strength of bricks using the Silver Schmidt sclerometer. This curve considers existing recommended curves for the mechanical version of sclerometers, with the goal of diagnosing masonry structures built to high workmanship standards, such as viaducts and bridges. So far, the most frequently used tool dedicated to quick and noninvasive assessment of structures is a standard sclerometer (Schmidt hammer), but mainly in the diagnosis of concrete structures [33]. Sclerometric tests using the Silver Schmidt electronic hammer enable verification of the homogeneity of the bricks used during the construction of the structure and a preliminary estimate of their compressive strength. Sclerometric tests carried out on brick structures with higher standards of workmanship and the quality of the material used are characterized by a smaller spread of results, which has a beneficial impact on the accuracy of estimating the compressive strength of the wall. There are several documents describing the sequence of proceedings for the sclerometric procedure, mainly regarding the testing of concrete structures, such as the American standard ASTM C805 [34], Polish PN-EN 12504-2 and others [33,35], or the Chinese JGJ/T 23-2011 [24]. So far, no standards have been developed for testing masonry structures, but RILEM Instruction TC 127-MS.D.2 [36] is available, describing a procedure for masonry structures that differs significantly from the one described for concrete structures. Inconsistent results may arise from the lack of standardized testing protocols for masonry structures. A standardized testing procedure must be created and followed to guarantee data accuracy and comparability. The main difference lies in the repeated measurement at a single point, as opposed to concrete testing where measurements are taken at intervals of 20–25 mm. Selecting the test site is crucial to ensure the following conditions: the tested brick, as well as the surrounding bricks and mortar, must be free of cracks and dry. Tests are not performed on edge bricks. Before starting the actual measurement, the hammer pin is placed perpendicularly to a clean and smooth surface and 3–4 strokes are made to better seat the pin head. Then, without removing the pin from the surface, 10 strokes are performed, and the resulting rebound R values are recorded. From these values, the five highest readings are selected. The average value of these selected rebounds, along with the standard deviation, constitutes the result for the given measurement location. Conversion of the obtained results into the compressive strength of bricks is only possible if a minimum of four destructive tests are carried out simultaneously. The number of tests carried out within one area of the structure may depend on the variability of the results obtained during the test, hence a greater number of measurements are required. The outcomes' variability is not the only factor determining the number of tests. Adhering to a specified testing process should ensure the consistency and dependability of the results. In construction practice, tests are most often carried out according to the rules adopted for concrete structures [37].

One of the first publications devoted to the use of a sclerometer to estimate the compressive strength of bricks was the above-mentioned study by Olek et al. [27]. Unlike the current methodology of testing whole bricks, the past standard provided for determining the f_B value on a sample made of two halves of a brick bonded together with cement mortar. The error estimated in the studies was 34%.

Conversion curve equation according to [7], Formula (3) for previous (information above):

$$f_B(R) = 0.031R^2 - 1.164R + 13.418 \tag{3}$$

The results of research on historical bricks (18th and 19th century) and those manufactured today with the methods used in the past (hand-formed bricks) were published by Egermann [38]. Based on the correlation between the compressive strength of bricks determined according to German standards and the rebound number recorded with the Schmidt hammer, he proposed his own conversion curve. This curve's abscissa represents the average rebound number R value, determined for the X direction (measurement on the headers) and the Z direction (measurement on the stretchers).

Conversion curve equation according to [38], Formula (4):

$$f_B(R) = 0.04{R_{xz}}^2 - 0.55R_{xz} + 13.6 \quad (4)$$

In these tests, a difference was recorded in the readings taken on headers and stretchers, which is related, among other things, to the anisotropic properties of ceramics.

R. Schrank recorded rebound numbers for the 19th-entury bricks in walls made of solid ceramic bricks in lime mortar [37]. He also determined the strength of bricks on the basis of destructive tests according to the DIN 105 standard. The same methods of increasing the confidence level are used in the sclerometric assessment of concrete strength when it is not possible to calibrate the base curve measured in destructive tests.

Conversion curve equation according to [37], Formulas (5) and (6):

$$f_B(R) = 0.00103R^3 - 0.058R^2 + R \quad (5)$$

$$f_B(R) = (0.00103R^3 - 0.058R^2 + R)/1.4 \quad (6)$$

The conversion function developed by Brozovski [39] was based on the results of the research in which a light-type mechanical Schmidt hammer was used. Nondestructive testing was carried out in accordance with EN 12504-2 [33] and CSN 731373 [40] standards, and the compressive strength was determined in accordance with EN 772-1 [41]. Two types of solid ceramic bricks with different dimensions were used, 29 × 14 × 6.5 (cm) and 25 × 12 × 6.5 (cm). The curves were developed for an LB hammer with a rounded shank (dedicated specifically to the diagnosis of masonry structures). It is noteworthy that during the tests with the LB hammer, a determination coefficient R^2 value of 0.956 was obtained, whereas for the L hammer, it was only 0.756. This suggests the greater efficacy of the LB hammer in diagnosing masonry structures.

Tests conducted by Roknuzzaman [42] allowed for the development of two relationships dedicated to testing bricks in a horizontal position. Both curves were characterized by a high value of the R^2 coefficient, i.e., 0.96 and 0.92.

To summarize the content of the above provisions, it should be stated that a uniform method for assessing the compressive strength of bricks in situ has not been developed so far. All the aforementioned attempts could only approximate the strength of the bricks within varying degrees of error. This could be significant for structures already under stress, even if they appear massive. However, there is a fundamental problem here when the research concerns the assessment of wall parameters (strength and homogeneity of wall elements) in larger areas. Their results can only be considered highly approximate due to the scope (not only the method), but they nevertheless relatively indicate the precise differences in the examined wall surfaces and their structures. In situ tests should always be verified by destructive tests, which will not always be possible due to the limited number of samples to be selectively taken if such a situation is possible at all.

4. Materials and Methods

4.1. Materials

The supporting structure of the castle ruins currently consists of walls and vaults made of solid ceramic bricks with a strength ranging between ~5 and 15 MPa, made of sand and lime mortars, the strength of which was set at 0.5 to 1.0 MPa. Brick is the basic building material of walls here. It was highly degraded biologically and erosively as a result of repeated cycles of strong dampness and drying, as well as by aeolian factors where solid particles carried by the wind caused abrasion of its external surfaces [6,24]. During visual inspection of the walls and vaults, no whitening or salt efflorescence was found, which indicates that the facility was not strengthened or repaired using cement-based materials. Locally, biodegradation was found, mainly on the upper surfaces of the vaults, as a result of the penetration of the roots of the intensively growing vegetation there. However, in the capillaries, there was no water rising from the ground. The main causes of moisture were climatic factors (precipitation, wind, temperature) and the lack of protection of the facility from above. The authors of this article decided to carry out nondestructive tests of the walls in situ and t collect samples of the material for laboratory tests. It was also decided that dirty places would not be cleaned unless the existing coatings were harmful to building ceramics; however, it was considered necessary to carry out biocidal treatments on walls and vaults after removing organic materials and vegetation from them. During the tests, no harmful coatings were found on the bricks, and only those surfaces that were intended for subsequent strengthening were cleaned.

4.2. Methods

4.2.1. Testing the Strength of Bricks

As written above, part of the material tests were carried out directly on the site. First of all, they concerned the evaluation of the strength of the embedded bricks using a Schmidt sclerometer (Figures 6 and 7). The findings were compared with the results of destructive tests performed on samples taken at several locations overlapping with the locations of the nondestructive tests. Sclerometric tests confirmed the high structural and strength inhomogeneity of the bricks, which was a further impediment to carrying out subsequent structural protection. The strength of the bricks, determined by the nondestructive method, was determined based on the correlation between their strength and the number of sclerometer reflections using the regression curve formula, among others, for the "N" type hammer, developed by Prof. Leonard Runkiewicz and Eng. Wieslaw Rodzik [25]. Therefore, the compressive strength values obtained this way were considered only illustrative due to the considerable scatter of results, the scale of the object size, the heterogeneity of the material, and the inaccuracy of the testing methods. They did not give the expected unambiguous answer about the current state of the remains of the castle walls, so it was decided to leave the structure of the object unchanged and propose another form of protection.

Core drillings were taken and then tested in a strength press. The results are given in Table 1.

Table 1. Brick test results.

Sample No.	Sample Weight	Average Height/Length of the Prepared Sample	Cross-Sectional Area	Destructive Force	Compressive Strength
	[kg]	[mm]	[mm^2]	[kN]	[N/mm^2]
1	0.867	98/98	7543	37.8	5.0
2	1.387	154/98	7543	83.1	11.2
3	1.02	122/98	7543	39.4	5.3
4	0.795	98/85	7543	37.2	4.9
5	0.946	99/98	7543	40.3	5.5
6	0.835	89/89	7543	42.1	5.4
7	0.963	96/89	7543	60.4	7.2
8	1.234	121/92	7543	41.1	5.6
9	1.097	101/90	7543	68.0	6.1
10	1.125	124/93	7543	38.2	5.1

SCLEROMETRIC MEASUREMENTS no 1/11/2022

Object	The walls of the castle in Słońsk	Date of construction	15th century
		Test date	07 / 11 / 2022
Element	interior wall	Expected grade of bricks	15 MPa
		Schmidt sclerometer type: N Concerto	
Owner	City of Słońsk	Calculations according to (Polish) standards: PN-B-06250: 1988 "Ordinary concrete" PN-74/B-06262 "Non-destructive testing of concrete structures" INSTRUCTION ITB 210/1977	

No.	Angle α	Readings Li									Average reading Li α	Angular correction ± ΔL	Average reading brought down Li	$(Li - \overline{Li})$	$(Li - \overline{Li})^2$
		1	2	3	4	5	6	7	8	9					
1	0	38	36	36	36	35	36	34	35	35	35,7	0,0	35,7	5,5	30,2500
2	0	34	34	30	34	36	36	34	35	35	34,2	0,0	34,2	4,0	16,0000
3	0	17	24	22	~~14~~	23	25	24	25	23	22,9	0,0	22,9	-7,3	53,2900
4	0	28	26	30	32	30	32	28	26	29	29,0	0,0	29,0	-1,2	1,4400
5	0	42	40	40	42	41	44	43	42	42	41,8	0,0	41,8	11,6	134,5600
6	0	30	29	29	28	26	25	24	29	28	27,6	0,0	27,6	-2,6	6,7600
7	0	22	21	20	23	23	22	18	21	21	21,2	0,0	21,2	-9,0	81,0000
8	0	30	31	35	30	34	~~42~~	32	30	28	31,2	0,0	31,2	1,0	1,0000
9	0	27	24	24	29	27	26	27	26	26	26,2	0,0	26,2	-4,0	16,0000
10	0	32	32	32	32	31	32	32	33	~~48~~	32,0	0,0	32,0	1,8	3,2400
Brick age: > 1000 **days**											Σ ➡		301,8	-0,2	343,5400

Angle α means the position of the Schmidt hammer during measurement.

$\overline{L}$ = *30,2*

S_L = *6,18*

v_L = *20,46 %*

Calculation factors:

The age of wall	c_t =	*0,60*
Bricks humidity	c_w =	*1,00*
Bricks humidity type:	**Powietrzno - suchy**	
Hypothetical coefficient:	C_h =	*1,00*

Brick quality indicators:

k_R = *0,01* v_R = *60,48 %*

$\overline{R}$ = *52,8 MPa*

R_{min} = *0,4 MPa*

S_R = *31,94 MPa*

Regression curve parameter
a = 0,30500
b = -11,42000
c = 131,60000

Guaranteed strenght**R**	= *0,4 MPa*
Momentary strenght	= *52,8 MPa*
Real class	= *< B 5*
Bricks quality	= *low*

Figure 6. Brick test results obtained with an N-type sclerometer (own study).

(**a**) (**b**)

Figure 7. (**a**) Testing the compressive strength of bricks using an N-type sclerometer; view of a sample (**b**) (own study).

4.2.2. Testing the Degree of Moisture of Bricks and Salt Content

The subsequent tests were carried out to assess the level of moisture and salinity of the masonry structures.

A commonly used measure of moisture content in building partitions, including walls made of ceramic bricks, is their mass (absolute) humidity. It is described by the percentage ratio of the weight of water contained in the material to its dry weight, as shown in Formula (7).

$$w_m = \frac{m_w - m_s}{m_s} 100\% \tag{7}$$

where

w_m—mass moisture, w %
m_w—wet weight of wood (sample), in g (kg)
m_s—mass of wood (sample) dried to solid mass, in g (kg)

Mass moisture measurement was performed using the Protimeter MMS2 meter (Protimeter, Crown Industrial Estate Priorswood Road Taunton TA2 8QY, UK (Figure 7). These tests were verified using the laboratory dry-oven test method. The results are presented in Table 2 [43].

Table 2. Results of moisture level tests for selected extreme cases of brick moisture.

Sample No.	Moisture	Mark	Moisture Meter Reading
1	5.56%	moderately moist	≤650
2	15.98%	wet	~999
3	16.82%	wet	~999

The moisture content of the bricks and the salt content in them were measured in a total of 50 samples. Only three test results showing the extremes of the measured quantities are included in Table 2, while Figures 8 and 9 (for its readability only) includes a graph showing the measurement results obtained at 20 measurement points. Accordingly, the salt content of samples 1, 2, 3 (according to Table 2) was determined.

Figure 8. Assessment of the moisture level of walls using the radio and laboratory dry-oven test method (photos by the authors).

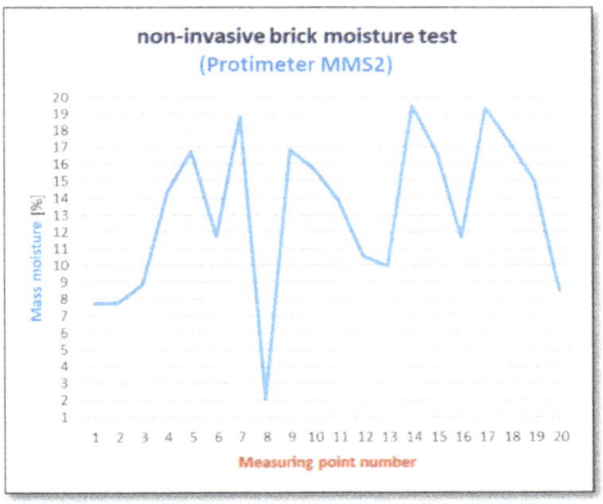

Figure 9. Results of in situ mass moisture measurements using a meter (measurement depth up to 20 mm).

The graph in Figure 8 shows significant differences in the moisture content in bricks at different measurement points. Until the roofing over the building structure was completed, the walls and vaults showed a very high level of moisture, so they were considered wet.

The moisture distribution on the wall surfaces was not uniform. The highest moisture levels were observed in the upper parts of the walls where they directly absorbed atmospheric precipitation. Conversely, the lower parts exhibited the least dampness, with a moisture content of less than 6% (7). The roof made in the spring cut off the rainwater supply, and in the summer, it led to their rapid drying, which was intensified by the strong ventilation of the building without windows and doors. This stabilized and evened out the moisture content in the walls to a level not exceeding w_m = 6% (7). However, this situation had an impact on the condition of the heavily damp bricks, which partially lost their compact structure, reducing their active load-bearing cross-sections. The process of natural drying of the wall will continue for several years and it is recommended to continue this method of drying the facility.

As part of the tests carried out, the level of salt content was determined using the chemical indicator method (Figure 10), the test results are presented in Table 3.

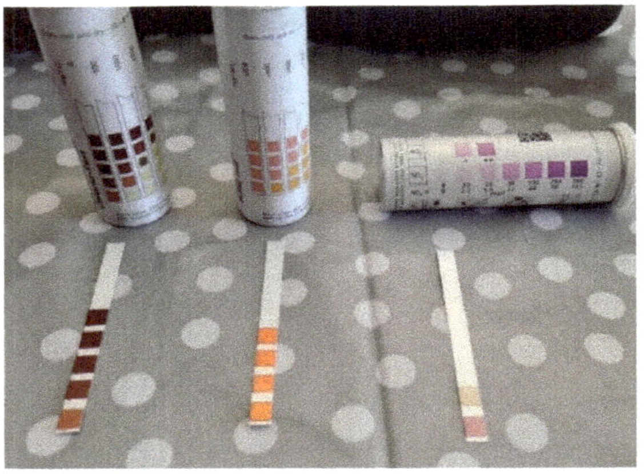

Figure 10. Assessment of salt content carried out on 20 samples taken from various places in the building (photo by authors).

Table 3. The determined salt content in the collected brick samples (own study).

Type of Salt	Determined Values (% of Mass) Sample No. 1	Sample No. 2	Sample No. 3	Compartment Qualifying (% of Mass)	Mark
nitrates NO_3	0.005%	0.005%	-	<0.100	low
sulfur SO_4	0.2%	0.2%	-	<0.500	low
Chlorides Cl	0%	0%	-	<0.500	low

Since the level of nitrates, sulphates, and chlorides was low, these compounds were excluded as the cause of the damage, which confirms the previously stated thesis that the direct cause was the variability of weather conditions and the lack of a roof over the facility.

Another problem and threat to the ruins of the building, apart from the heterogeneity of the wall structure, was the deformation of the high 19 m walls in three directions. The deflection of walls from the vertical (including those in the shape of an arc) reached up to 40 cm and, locally, even up to 1.0 m.

It was useful to conduct in situ and laboratory tests to assess the condition of the entire structure.

The authors of this article emphasize the importance of material research necessary to maintain the remains of historic buildings that are in a state of technical and functional ruin. The ongoing research of the medieval castle in Słońsk is intended to serve both the preservation of the assessed materials and its remaining ruins for the purposes of further historical searches, archaeological exploration, and public access, constantly increasing knowledge about them. Within the framework of this article, the case of preservation of the remains of the chimney wall located in the northern corner of the building (Figures 5 and 11) is selectively described, bearing in mind the incompletely recognized parameters of the masonry elements due to their heterogeneity in such a sizable volume of the building (about 22,500 m^3). Attempting any reinforcement, additions, or repointing of the object and its elements would lead to a reduction of its historic character. Therefore, the results of the research provided a partial answer to the issue of further possible although risky treatment. This is how material and environmental research should be directed, and it should not be an end in itself.

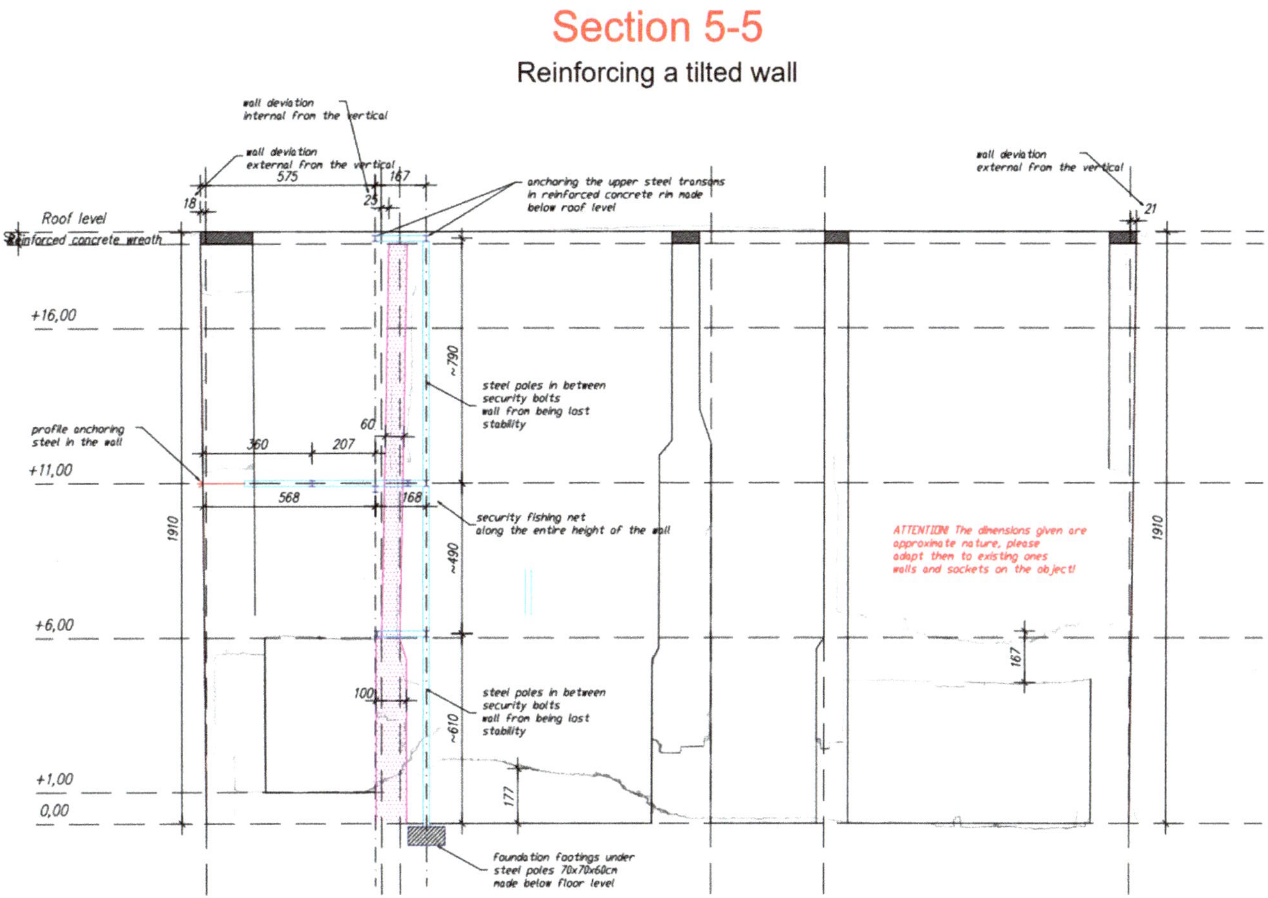

Figure 11. Cross-section through the chimney wall and its front view (own study).

The wall in question, intact, was to remain in the object as a witness to the older and recent history of the castle, depicting the turn of its fate to the present day, and the results of the material tests carried out were to assess the possibility of allowing such exploitation. The investigated case is not a duplication of similar standard research but a way of using it for practical as well as research purposes for a specific object, which, thanks to it, was admitted to exploitation bearing in mind the danger that may be posed by the incomplete building materials that make it up. The main objective of this research was to establish the parameters of medieval materials degraded by being left in an environment to which they were not adapted. Their age, fire, and exposure to the external environment without any protection contributed to bringing them to such a state. In addition, the castle's vaults were subjected to loads of rubble, organic embankment (soil), and vegetation whose roots and moisture had a very negative impact on the structure of the walls and the bricks and mortar themselves.

As part of the analysis and calculations carried out, it was determined that in the extreme case, even after securing the wall structures, there is a possibility of their damage during further operation, including in the presence of tourists. In view of the above, such a state of danger was provided for in the construction solutions. It was proposed to make an independent structure to protect the surroundings from the effects of possible dehiscence and fall of loose bricks and even whole fragments of the wall while respecting the authenticity of the monument. The research and subsequent analysis of the structure in a highly deformed state will allow to keep the monuments—ruins—in the current authentic state for them, not meeting the criteria of bearing capacity and stability required by current regulations and standards [18].

The idea behind this solution was to capture the damaged wall at three levels and protect it against loss of stability by anchoring it to a wall where stability and load-bearing capacity were beyond doubt. The project assumed the possibility of fragments of the wall breaking off in the highest part of the wall and the possibility of them falling to the lowest level, hence the designed steel structure was secured with a steel mesh. This solution took into account the low-strength class of bricks and mortars, their deep losses, and the lack of proper bonds between the masonry elements. In the proposed solution, platforms made of 25 mm thick OSB (Oriented Stand Board) boards were introduced on two levels, the task of which was to dampen the impact of falling debris on two levels, assuming the possibility of their subsequent destruction (starting from the first and then the second platform) in the event of the disaster of the wall in question. The principle of local strengthening of the above-mentioned chimney wall is shown in Figure 11, which was to involve injecting gaps in the joints between the bricks and strengthening the cracked and bulging parts of the walls with composite meshes (FRCM). The use of carbon meshes applied on mineral matrices did not take away from the authentic appearance of the monument because, in the past, its walls were entirely covered with plaster.

The structure described above was actually made with minor modifications, adapting it to the possibility of installation in the immediate vicinity of the deformed walls. Its final and implemented version is shown in Figures 12 and 13. Currently, this facility is partially open to the public.

As part of the measures available to the current owner of the building, the vaulted lintels were also secured using a method similar to that described above, i.e., without any rebuilding or strengthening. The aim of such actions was to limit the loss of the structure's authenticity as much as possible.

As a result of covering the ruins with a temporary roof, the facility began to dry out quickly, which had an impact on the building ceramics, especially the vaults, which, after removing unnecessary ballast and cleaning, turned out to be severely deformed structures with local, but deep, defects. Moreover, as a result of drying, the bricks lost their structure and their bottom surfaces became detached to a depth of up to half of their thickness. Therefore, the planned method of strengthening them on both sides with composite meshes on mineral mortars could only be applied locally and only in less damaged structures.

Some of the vaults required additions (Figures 14–16). This was achieved by reproducing the original, using the recovered bricks.

As part of the rescue and security works, biological contamination was neutralized with biocides in the form of an alkonium-free chloride solution (pH: approx. 7.5) dedicated to removing algae, fungi, lichen, and moss spores from the surfaces of the mineral building materials. This measure was used in places where organic materials had previously accumulated on the vaults and the ground floor at their junction with the walls, after their prior removal. Agents based on silicic acid esters were used to consolidate the weakened structure of the bricks. In places where the application of strengthening composite materials was planned, the load-bearing capacity and adhesion to the surfaces of ceramic walls and vaults were checked using the pull-off method. The peel strength of the test discs should not have been less than 1.5 MPa.

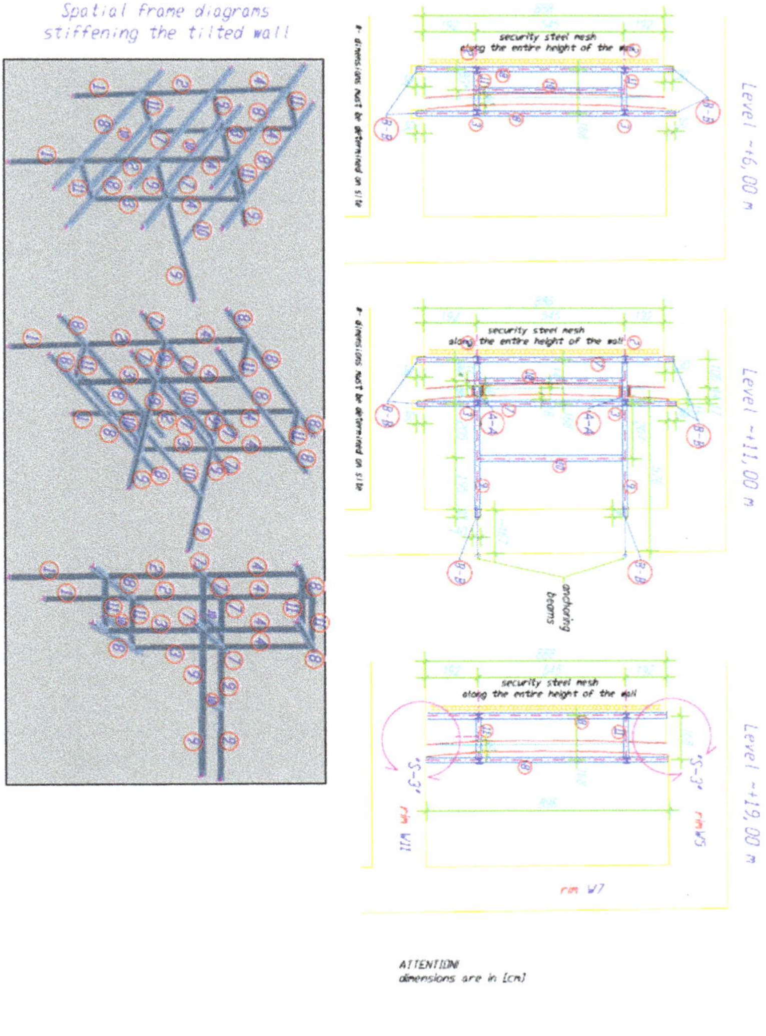

(a)

Figure 12. *Cont.*

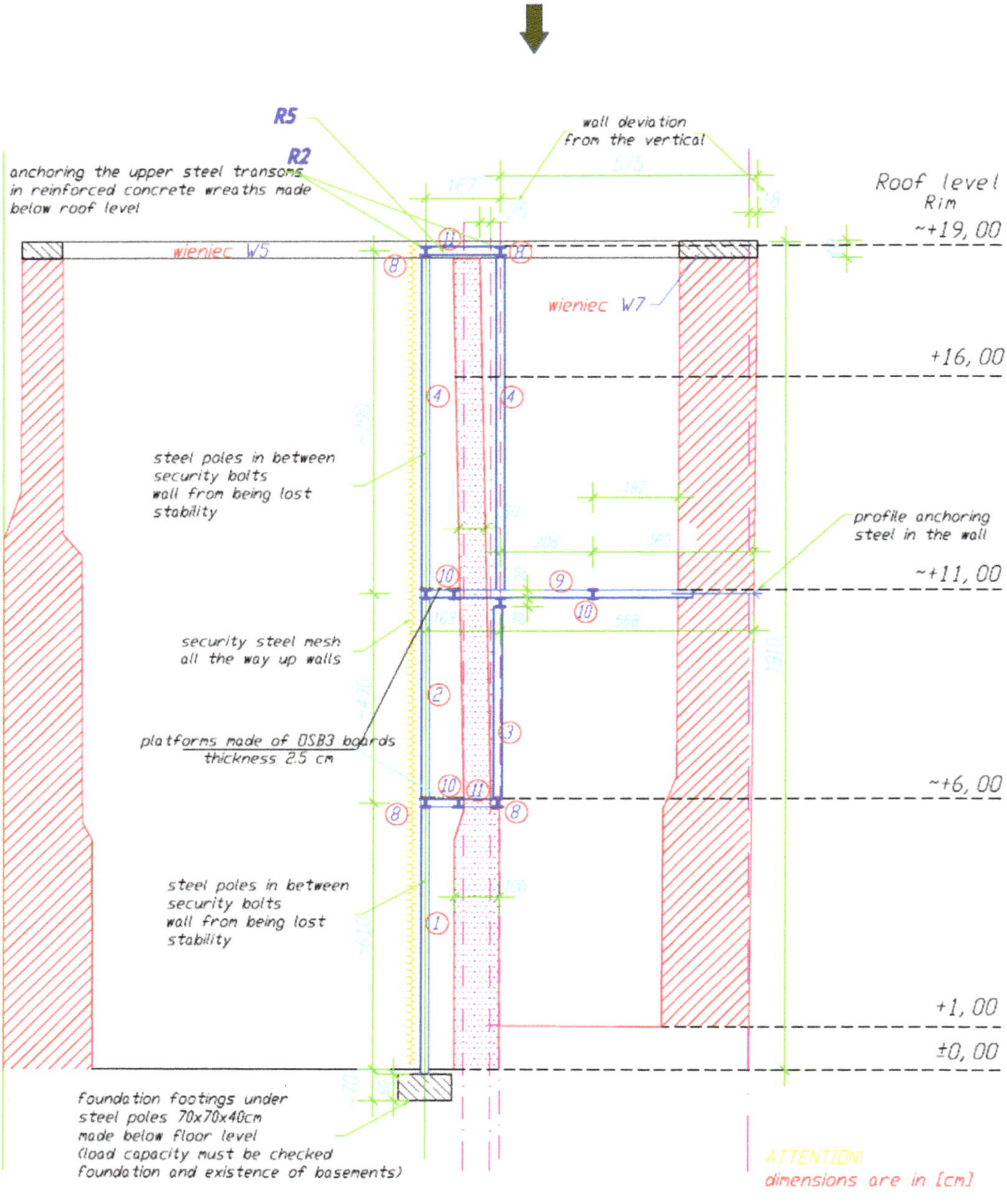

(**b**)

Figure 12. Diagram of an independent structure strengthening the chimney wall (**a**) frames, (**b**) cross section (own study).

(a) (b) (c)

Figure 13. Proposal for strengthening the chimney wall: (**a**) independent steel structure (two platforms intended to transfer the load from falling debris are marked in red); (**b**) places where bricks are joined by the injection of joints (yellow); (**c**) places where strengthening composite meshes (FRCM) are applied—green (photos by authors).

Figure 14. Current view of the strengthened wall and the strengthening structure (photos by authors).

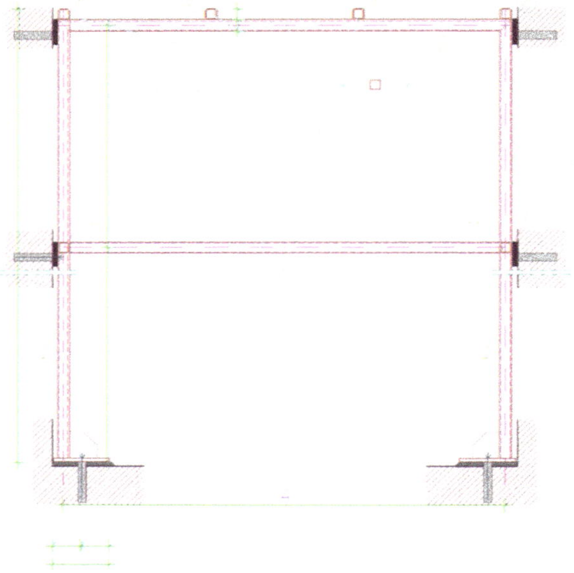

(a)

(b)

Figure 15. Securing the arches of openings in walls: (**a**) design; (**b**) execution (own study).

(a)

(b)

Figure 16. Fragments of the vaults above the ground floor of the castle: (**a**) view of the damaged bricks from below; (**b**) filling in (photos by authors).

The most damaged "chimney" wall was strengthened and protected by:

- Preliminary ad hoc stamping on both sides;
- Cleaning the floor of the rooms and setting up scaffolding;
- Construction and assembly of a steel reinforcing structure and equipping it with light platforms on two levels (Figures 10–12);
- Placing OSB3 boards on platforms to reduce the effects of the impact of a falling wall fragment—they may break, which will allow the rubble to slide more gently to the lower level and the ground floor;
- Filling larger gaps between vault bricks with expansive mortars based on lime, cement, and trass;
- Introduction of reinforcements on damaged wall surfaces using C-FRCM meshes on inorganic mortars; *
- Removal of temporary stamping structures and scaffolding.

* Damaged wall structures were surface-reinforced with carbon fiber meshes on inorganic mortars based on the Ruredil C-MESCH GOLD 84/84 system (Visbud-Projekt Sp. z o.o. Wrocław, Poland). A layer of inorganic matrix based on Ruredil X Mesh M25 pozzolans (Visbud-Projekt Sp. z o.o. Wrocław, Poland) for masonry substrates with a thickness of 3 mm was applied on the stripped, cleaned, and moistened brick substrate, after prior checking of the adhesion of future composites to the substrate using the pull-off method (the pull-off strength of the discs exceeded 1.5 MPa). In the next stage, the Ruredil X Mesh C10 (Visbud-Projekt Sp. z o.o. Wrocław, Poland) mesh was "embedded" and a second layer of Ruredil X Mesh M25 mortar (Visbud-Projekt Sp. z o.o. Wrocław, Poland), also 3 mm thick, was applied (Figures 11 and 12). One-sided or double-sided reinforcement or strengthening of the vaults was made as described above after injecting the gaps in the joints between the bricks.

5. Results

A wide range of tests carried out on the facility and, as a result, the proposal and implementation of very low-invasive interventions in the structure of the walls made it possible to stop the degradation processes of this structure and its elements while maintaining them in a visually unchanged (authentic) state. Currently, it has been monitored for 3 years, and at the same time, it has been made available to the public for viewing (Figure 17).

(a) (b)

Figure 17. Ruins of the castle in Słońsk: (**a**) initial condition; (**b**) current condition (photos by authors).

Test methods for valuable structures and building materials should be selected so as to pose the least possible threat to unique, historical structures while maintaining the highest degree of their authenticity (Table 4).

Table 4. Brief description of methods of dealing with the tested object.

Name	Description
1. Preparatory work	➢ Cutting off the supply of further portions of moisture from the walls of a facility that has not been protected for 50 years (roofing and natural drying time); ➢ Literature review in the field of research methods used and currently being developed; ➢ Historical inquiry regarding the past of the monument and changes that have been made to it, including the period of neglect over the last 50 years; ➢ Preliminary inspection of the entire facility; ➢ Preparing the remains of the castle structure for inventory—enabling safe access to it. ➢ Inventory of structures—manually and digitally (3D scanning);
2. Laboratory tests	➢ Selection and indication of places to take samples of bricks and mortars and then collecting them; ➢ Destructive compressive testing—bricks and bricks containing mortar fragments—support joints (these bricks were previously subjected to nondestructive testing using a sclerometer to determine the hypothetical coefficient); ➢ Determining the water content in bricks; ➢ Determination of salt content in bricks; ➢ Determination of brick absorption;
3. "In situ" brick testing	➢ Visual assessment of damage; ➢ Sclerometric tests of the compressive strength of bricks. ➢ Assessment of the degree of moisture using a nondestructive method; ➢ Measurement of erosion losses (including mortars in joints);
4. Analysis of the study results	➢ Assessment of the strength of individual samples and the dispersion of results. ➢ Assessment of the size of net section losses in bricks and mortars; ➢ Assessment of the load-bearing capacity of the wall; ➢ Selection of a method for structural bonding of bricks heavily degraded by the external environment.
5. Determining the amount of load that may be imposed on existing walls after taking into account significant differences in the load-bearing parameters of the bricks and mortar losses in the joints, as well as taking into account load imperfections.	
6. Developing rescue and repair methods for damaged masonry structures.	

For nondestructive tests of the compressive strength of bricks, Formula (1) for the regression curve of L. Runkiewicz according to [25] was used, which gave values close to the results obtained in destructive tests.

6. Discussion and Conclusions

The problem of saving monuments, or more precisely the materials from which they were made, is very complex despite appearances indicating the simplicity of their construction. Often, in addition to the knowledge of the principles of statics and the properties of the materials used, it is also required to take into account the effects of physical processes that may occur inside the partitions. There are no two identical buildings or identical structures or materials, hence any interference with a historical building should take into account not only protecting it against further degradation, or even destruction, but also preserving and exhibiting the solutions used in it, as well as the effects of the solutions proposed in it, which should also be monitored in their future operation. In this article, the authors describe a case that occurs commonly and often leads to the reduction of national heritage resources if such a problem is not noticed and treated appropriately at the right time. More and more often, the so-called permanent ruin is mentioned, i.e., something that not only does not have the same splendor as the original but is already in a state of "technical agony" but still constitutes a valuable cultural, scientific, and historical value.

This article is a brief summary of the possibilities of dealing with immovable monuments, protected by law, for which the combination of field, laboratory research, and analysis of the structure gives the most feasible picture of their preservation. The authors, using proven and pioneering research methods, developed a model for saving a highly damaged masonry structure, proposing an entirely different approach to preserving historically and scientifically valuable buildings/structures by leaving them unchanged or even damaged. The only change here was the introduction of lightweight, external, low-exposure structures in relation to the original to be protected, ensuring that they remain statically balanced and protected from danger to themselves and their surroundings. There are no methods that will unequivocally be able to assess the properties of embedded materials. This is due to the not yet fully developed testing techniques, the range of their capabilities, and above all, sizable areas (volumes) of the tested objects with heterogeneous structures. Not always and not everywhere will it be possible to successfully implement proper diagnostic techniques and traditional repair methods regardless of the type of material and technology used. This article is, in a sense, a case study; nevertheless, it can be successfully applied to other similar situations where urgent salvage work is required, which, due to the extent of the degradation present or the threat they may pose to themselves and their surroundings, is not possible to implement in the traditional manner. At present, the site, left in a state of safe (permanent) ruin, allows for the research to continue and, at the same time, is open to the public, serving as a tourist attraction. The targeted and ad hoc safeguards applied thanks to the knowledge of its properties allow for historical and archaeological research to continue in it. Since the walls and vaults are still drying out naturally and this process will continue for several more years, along with the changes in the structure of the bricks, monitoring these materials and structures by measuring moisture content and checking for any deformation continues. In such situations, it should not be limited to conducting material tests alone in isolation from their role in the construction of the structures they form, especially in the case of large-scale structures. In such cases, the heterogeneity of the parameters and the state of preservation of the materials—here, bricks and mortar and their interrelationships—determine the strength and stability of what they were used for. The authors of this article constantly carry out monitoring of the masonry structure, taking into account changes in the parameters of bricks and mortars over time due to their drying out after securing the castle with a roof. Such an activity is an excellent testing ground for the change in operating conditions to the already favorable indoor climate and progressive degradation of the materials originally examined, mainly due to their drying out and changes in structure. Such research is still being conducted and the results will be the subject of a separate publication.

The literature review along with the research and practical experience of the authors of the article confirms the statement that there are no universal research methods, especially if they concern centuries-old building materials. In addition to the assessment of physical and strength parameters, there remains the historical aspect of the technology and conditions of their (old buildings) execution. Decisions made on the basis of research carried out contain a certain amount of risk, which is particularly dangerous for unique materials and structures made from them (historical materials). Hence, the results of all tests should be treated as initial material for in-depth analyses, both in terms of static strength assessments and those supported by historical queries. The authors of this article have determined the preliminary parameters of the bricks in terms of their strength, structural integrity, and harmful moisture. The fourth parameter of the research in question was monitoring the facility with a simultaneous analysis of the changing environmental humidity conditions (after roofing the ruins) and their impact on changing the properties of bricks and mortars losing the above-mentioned structural integrity. This condition of the bricks also influenced the results of nondestructive tests of the bricks where their top layer with a loose structure influenced the reading of the size of the Schmidt sclerometer reflections and, thus, lowered the read compressive strength of the bricks. The loose structure of the bricks also eliminated the initially assumed method of strengthening the vaults by applying composite materials to their bottom surfaces. The conducted research facilitated the development and selection of strategies aimed at preserving the original structural integrity and external aesthetics of the building while ensuring the necessary level of reliability. As part of the safety and repair works, three different methods of strengthening and securing the structure were introduced, which was adequate to the scope of the damage and their location. The authors of this article also considered the potential challenges associated with predicting the behavior of the structure, particularly concerning the loss of cohesion in unexamined areas of the walls. The efficacy of the diagnoses and chosen solutions will depend on continued monitoring and observations of the structure over time.

Apart from the scientific and technical aspects, there is also a social aspect. The materials that are the subject of this article (brick, mortar) constitute a structure for which an appropriate role has been designated, and its parameters preserved so far are to ensure the durability and safety of the remains of the facility. The article not only describes a case study but also covers the problem in much greater detail, going beyond the framework of typical laboratory tests and the creation of statistical curves, which are an end in themselves. We are dealing here with a valuable historical object for which the authors of this article accepted responsibility. The article indicates methods of handling severely degraded structures caused by damage to their components, which are periodically exposed to unique loads over a long period of nearly 50 years, which are parallel to the standards used in the case study. This is not a typical load case associated with degraded materials constituting the building structure, and the proposed methods based on commonly used technologies are not conventional. This article shows a way to preserve the existing condition (and even appearance) of materials without using invasive techniques and materials that change their image and character. An inherent element of such tests is static and strength calculations, which are not attached to the article due to their extensiveness. Nevertheless, in situ tests, verified by random laboratory tests, have indicated the possibility of further operation of the damaged facility. The measurement of humidity and the examination of the absorbability of the bricks allowed for a clear conclusion that these were not masonry elements adapted to external exposure without additional protection, i.e., plasters. This conclusion concerned both the external and internal walls of the castle. No two objects are the same even though they could be made of similar materials, by the same work team, and in identical circumstances. Hence, in such cases, there are no universal research methods that can solve every problem. The authors pointed to an unusual form of solving the problem, which may be applicable in other similar cases but only after prior assessment of the degree of degradation of the object and the threat it may pose. As it is written in the article that nondestructive methods of estimating the strength of bricks and mortars are still in the research stage, most such structures are not able to wait until the end of the research and testing stage. What we are dealing with here is a variety of masonry elements in one historic structure at a risk of loss. The research carried out was of an implementation nature, which is currently being verified through observation and further analyses in terms of possible new damage, deformations, and durability of the solutions used. This masonry structure is the target of further interest, observation, and research.

The research carried out allowed for the assessment of the state of preservation of the remains of walls and vaults of a valuable medieval monument with a height of 19 m and a volume of 22,500 m^3, which consequently saved ~8000 m^3 of walls and 210 m^2 of the remains of vaults made of solid ceramic bricks in lime mortar. In this case, the structures preserved in their original state should be considered as the starting research material for further investigations and monitoring of the facility, especially in terms of the effectiveness and durability of the structural bond of the bricks that have lost their cohesion as they dried out due to the cutoff of the "moisture supply". Continuation of geodetic measurements will allow for the assessment of the stability of the entire masonry structure and, thus, the analysis of its degree of safety as a function of time, which should be helpful in situations that limit the possibility of taking samples for destructive testing or even in situ testing.

Nowadays, we hear more and more often about leaving historical objects in the so-called "permanent ruin", hence the research described in this article may prove very useful when making such risky decisions. When deciding to leave historical objects in their current condition, the European Monument Protection Services refrains from trying to restore their original appearance and function because this would involve interference that would reduce their historical value and authenticity. On the other hand, masonry structures left in this form may not fully meet the safety requirements imposed by currently applicable regulations and standards. Researchers of objects and of building materials should not only analyze and evaluate the obtained research data and calculation results but also develop rescue and repair methods, assuming responsibility for the decisions made and, in such cases, protecting cultural heritage.

Author Contributions: Conceptualization, D.B. and K.S.; methodology, D.B.; software, D.B.; validation, D.B., K.S., and A.G.; formal analysis, K.S.; investigation, D.B.; resources, K.S.; data curation, D.B.; writing—original draft preparation, D.B.; writing—review and editing, K.S.; visualization, A.G.; supervision, D.B.; project administration, D.B. All authors have read and agreed to the published version of the manuscript.

Funding: This research received no external funding.

Institutional Review Board Statement: Not applicable.

Informed Consent Statement: Not applicable.

Data Availability Statement: Data are contained within the article.

Conflicts of Interest: The authors declare no conflicts of interest.

References

1. Bajno, D. *Rewitalizacja Konstrukcji Budowlanych w Obiektach Zabytkowych*; Wydawnictwa Uczelniane Uniwersytetu Technologiczno-Przyrodniczego: Bydgoszcz, Poland, 2013.
2. Menzel, C.A. *Der Steinbau. Handbuch Fuer Architekten, Bauhandwerker und Bauschueler*; Karlsruhe, J., Ed.; Blelefeld's Verlag: London, UK, 1885.
3. Bungey, J.; Millard, S.; Gratham, M. *Testing of Concrete in Structures*; Taylor&Francis: London, UK; New York, NY, USA, 2006.
4. Carino, N.J. *Nondestructive Test Methods, Concrete Construction Engineering Handbook*; CRC Press: Boca Raton, FL, USA, 1999.
5. Davis, A.G. The non-destructive *impulse* response test in North America: 1985–2001. *NDTE Int.* **2003**, *36*, 185–193. [CrossRef]
6. Drobiec, Ł.; Jasiński, R.; Piekarczyk, A. *Diagnostyka Konstrukcji żelbetowych. Metodologia, Badania Polowe, Badania Laboratoryjne Betonu i Stali, v. 1*; Wydawnictwo Naukowe PWN: Warszawa, Poland, 2010.
7. Kozak, R. *Budownictwo Betonowe. Tom VIII. Badanie Materiałów, Elementów i konstrukcji*; Arkady: Warszawa, Poland, 1970.
8. American Concrete Institute. *American Concrete Institute Report ACI 228.2R-98. Nondestructive Test Methods for Evaluation of Concrete in Structures*; ACI: Farmington Hills, MI, USA, 1998.
9. Aliabdo, A.A.E.; Elmoaty, A.E.M.A. Reliability of using nondestructive tests to estimate compressive strength of building stones and bricks. *Alex. Eng. J.* **2012**, *51*, 193–203. [CrossRef]
10. Brencich, A.; Sterpi, E. Compressive Strength of Solid Clay Brick Masonry. Calibration of Experimental Tests and Theoretical Issues. Structural Analysis of Historical Constructions. 2006, pp. 757–766. Available online: https://www.researchgate.net/publication/281109671_Compressive_Strength_of_Solid_Clay_Brick_Masonry_Calibration_of_Experimental_Tests_and_Theoretical_Issues (accessed on 20 April 2024).
11. Debailleux, L. Schmidt hammer exposure dating for brick masonry. *Geochronometria* **2020**, *47*, 54–62. [CrossRef]
12. Bajno, D.; Bednarz, Ł.; Nowak, T. Damp proofing of heritage buildings located in historical town centres. In Proceedings of the SAHC2014—9th International Conference on Structural Analysis of Historical Constructions, Mexico City, Mexico, 14–17 October 2014; pp. 1–7.
13. Bednarz, Ł.; Bajno, D. Remains of urban heritage defence structures—Conservation, monitoring and use. In Proceedings of the Sahc2014—9th International Conference on Structural Analysis of Historical Constructions, Mexico City, Mexico, 14–17 October 2014; pp. 1–12.
14. Domasławski, W.; Kęsy-Lewandowska, M.; Łukaszewicz, J. *Badania nad Konserwacją Murów Ceglanych*; Uniwersytet Mikołaja Kopernika: Toruń, Poland, 2004.
15. Crocci, G. *The Conservation and Structural Restoration of Architectural Heritage*; WIT Press: Southampton, UK, 1998.
16. Bajno, D.; Grzybowska, A.; Gajewski, J. Problems with Maintaining in Required Technical Condition and Revitalization of Medieval Defense Fortries. In *Key Engineering Materials*; Trans Tech Publications Ltd.: Stafa-Zurich, Switzerland, 2019; Volume 817, pp. 665–672. [CrossRef]

17. Jasienko, J.; Bednarz, Ł.; Bajno, D. Conservation of the structure and materials of historic masonry walls. In *Konferencja MURICO 2014 (Włochy) Engineering Materials*; Trans Tech Publications: Stafa-Zurich, Switzerland, 2015; Volume 624, pp. 354–362. [CrossRef]
18. Eurokod 6 Projektowanie Konstrukcji Murowych Częśc1-1: Reguły Ogólne Dla Zbrojonych i Niezbrojonych Konstrukcji Murowych. Available online: https://integro.bs.katowice.pl/33005402682/normy/eurokod-6?contrast=default (accessed on 20 April 2024).
19. Construction Law Act: Ustawa z dnia 7 lipca 1994 roku—Prawo budowlane (jednolity tekst: Dz.U. z 2019 roku, poz. 1186 ze zmianami, tj. zmiany: Dz.U. z 2019 roku, poz, 1309, 1524, 1696, 1712, 1815, 2166, 2170). Available online: https://isap.sejm.gov.pl/isap.nsf/DocDetails.xsp?id=wdu19940890414 (accessed on 20 April 2024).
20. Act on the protection and care of monuments: Ustawa z dnia 23 lipca 2003 r. o ochronie zabytków i opiece nad zabytkami (jednolity tekst: Dz.U. z 2018 roku, poz. 2067, zmiany: Dz.U. z 2008 r. poz. 2245 i z 2019 roku poz. 730 i poz. 1696). Available online: https://isap.sejm.gov.pl/isap.nsf/DocDetails.xsp?id=wdu20031621568 (accessed on 20 April 2024).
21. Available online: https://polska-org.pl/9789739,foto.html?idEntity=7340286 (accessed on 29 March 2024).
22. Available online: https://www.google.com/search?client=firefox-b-d&q=zamek+w+s%C5%82o%C5%84sku#vhid=xi1TkQjvbeLxsM&vssid=l&ip=1 (accessed on 29 March 2024).
23. Gorek, M. *Skanowanie Laserowe 3D Obiektu Wraz z Opracowaniem Chmury Punktów*; Gorek Restauro Sp. z o. o.: Warszawa, Poland, 2022.
24. *JGJ/T 23-2011*; Technical Specification for Inspecting of Concrete Compressive Strength by Rebound Method. MOHURD: Beijing, China, 2011.
25. Runkiewicz, L.; Rodzik, W. Badania Nieniszczące Wytrzymałości Murowanych Obiektów Zabytkowych. Inżynieria i Budownictwo, No 2/1990.
26. Kubissa, J.; Strzelczyk, M. Nieniszczące badania wytrzymałości cegieł ceramicznych, "Inżynieria i budownictwo" No. 10-11/1987.
27. Matysek, P. Identyfikacja wytrztrzymałości na ściskanie i odkształcalności murów ceglanych w obiektach istniejących. Wydawnictwo Politechniki Krakowskiej: Kraków, Poland, 2014; pp. 103–105. Available online: https://repozytorium.biblos.pk.edu.pl/redo/resources/25701/file/suwFiles/MatysekP_IdentyfikacjaWytrzymalosci.pdf (accessed on 20 April 2024).
28. Instrukcja stosowania młotków Schmidta do nieniszczącej kontroli jakości betonu w konstrukcji, Instytut Techniki Budowlanej, Warszawa 1977. Available online: https://www.studocu.com/pl/document/politechnika-warszawska/budownictwo-ogolne/instrukcja-itb-210-instrukcja-stosowania-mlotkow-schmidta-do-nieniszczacej-kontroli-jakosci-betonu-w-konstrukcji-warszawa-1977/97134970 (accessed on 20 April 2024).
29. Filipowicz, P. Próba Zastosowania Młotka Schmidta do Badań Architektonicznych Ceglanych Murów. *Ochr. Zabyt.* **2009**, *1*, 38–42. Available online: http://cejsh.icm.edu.pl/cejsh/element/bwmeta1.element.desklight-a4a5324b-7fe0-46ab-a18f-d3e13b8e263e (accessed on 20 April 2024).
30. Mengistu, G.M.; Gyurkó, Z.; Nemes, R. The Influence of the Rebound Hammer Test Location on the Estimation of Compressive Strength of a Historical Solid Clay Brick. *Solids* **2023**, *4*, 71–86. [CrossRef]
31. Łątka, D. Rozprawa Doktorska "Wytrzymałość i Odkształcalność Ceglanego Muru—Ocena" na Podstawie Badań Nieniszczących i Małoniszczących", Politechnika Krakowska, Kraków 2023 (Strength and Deformability of Brick Masonry—Evaluation Based on Nondestructive and Semidestructive Tests). Available online: https://repozytorium.biblos.pk.edu.pl/resources/47383 (accessed on 20 April 2024).
32. Matysek, P.; Łątka, D. Comments on the application of the sclerometric method in the diagnostics of brick masonry. in Structural analysis of historical constructions. In Proceedings of the International Conference on Structural Analysis of Historical Constructions, SAHC 2012, Wroclaw, Poland, 15–17 October 2012; Volume 3, pp. 2471–2479.
33. *PN-EN 12504-2:2021-12*; Badania Betonu w Konstrukcjach—Część 2: Badanie Nieniszczące—Oznaczanie Liczby Odbicia. PKN: Oklahoma City, OK, USA, 2021.
34. *ASTM C805/C805M*; Standard Test Method for Rebound Number of Hardened Concrete. ASTM International: West Conshohocken, PA, USA, 2018. [CrossRef]
35. *UIC 778-3*; Recommendations for the Inspection, Assessment and Maintenance of Masonry Arch Bridges. International Technical Standard: Zdar nad Sazavou, Czech Republic, 2019.
36. RILEM Technical Committees. RILEM TC 127-MS.D.2: Determination of Masonry Rebound Hardness. *Mater. Struct.* **1998**, *31*, 375–377. [CrossRef]
37. Schrank, R. Materialeigenschaften historischen Ziegelmauerwerks im Hinblick auf Tragfähigkeitsberechnungen am Beispiel der Leipziger Bundwand. *Das Mauerw.–Z. Technol. I Archit.* **2002**, *10*, 201–206. [CrossRef]
38. Egermann, R. *Zur Nachträglichen Bestimmung der Mechanischen Eigenschaften von Mauerziegeln*; Ernst und Sohn: Berlin, Germany, 1992.
39. Brozovsky, J.; Zach, J. Evaluation of Technical Condition of Mansory Structures. In Proceedings of the 15th International Brick and Block Masonry Conference, Santa Caterina, Brazil, 3–6 June 2012.
40. *CSN 731373*; Nedestruktivní zkoušení betonu—Tvrdoměrné metody zkoušení betonu. TECHNOR print, s.r.o.: Hradec Králové, Czechia, 2011.
41. *EN 772-1*; Metody badań elementów murowych—Część 1: Określenie wytrzymałości na ściskanie. PKN: Oklahoma City, OK, USA, 2001.

42. Roknuzzaman, M.; Hossain, B.; Mostazid, I.; Haque, R.R. Application of Rebound Hammer Method for Estimating Compressive Strength of Bricks. *J. Civ. Eng. Res.* **2017**, *7*, 99–104. [CrossRef]
43. Jasieńko, J.; Matkowski, Z. Zasolenie i Zawilgocenie Murów Ceglanych w Obiektach Zabytkowych—Diagnostyka, Metodyka badań, Techniki Rehabilitacji. Wiadomości Konserwatorskie No. 14, Kraków 2003. Available online: https://www.semanticscholar.org/paper/Zasolenie-i-zawilgocenie-mur%C3%B3w-ceglanych-w-%E2%80%93-bada%C5%84,-Jasie%C5%84ko-Matkowski/286ec64bcb4fa52319fca931223da8ba681ef94c#citing-papers (accessed on 20 April 2024).

materials

MDPI

Article

Research on 18th-Century Building Structures in Terms of Static Scheme Changes

Monika Mackiewicz [1,*], Janusz Ryszard Krentowski [1], Kamil Ziminski [1] and Aldona Skotnicka-Siepsiak [2]

[1] Faculty of Civil Engineering and Environmental Sciences, Bialystok University of Technology, Wiejska 45E, 15-351 Bialystok, Poland; janusz@delta-av.com.pl (J.R.K.); kamilziminski@op.pl (K.Z.)

[2] Faculty of Geoengineering, University of Warmia and Mazury in Olsztyn, Heweliusza 4, 10-724 Olsztyn, Poland; aldona.skotnicka-siepsiak@uwm.edu.pl

* Correspondence: m.mackiewicz@pb.edu.pl

Abstract: The evaluation of the technical condition of historic buildings that have operated for several hundred years is a complicated issue. Even buildings that are in very poor condition must be checked and assessed in terms of their further repair, strengthening, or compliance with conditions that allow the facility to be safely operated. Most 18th-century buildings have not survived to this day retaining their original arrangements and structural elements. Renovations and repair work in the past were often carried out using materials of uncertain quality, with repair work of different qualities and without detailed analysis or methodology, based only on the experience of the former builders. In historic structures, the character of the work of individual structural elements has often changed due to significant material degradation, the poor quality of repair work, or the loss of adequate support. When load transfers change, internal forces are redistributed, and, as a result, the static scheme changes. This article presents an overview of identified defects affecting the change in static schemes in historical building structures built in the 18th century, using the example of a historic building with a large number of aforementioned defects. The process of assessing the technical condition of the facility is presented, in which non-destructive testing (NDT) methods were used. Detailed computational analyses were carried out for the wooden roof truss structure, which had partially lost its support.

Keywords: historic buildings; structural element damage; technical condition assessment; changes in the static scheme

Citation: Mackiewicz, M.; Krentowski, J.R.; Zimiński, K.; Skotnicka-Siepsiak, A. Research on 18th-Century Building Structures in Terms of Static Scheme Changes. *Materials* **2023**, *16*, 7689. https://doi.org/10.3390/ma16247689

Academic Editor: Francesca Ceroni

Received: 20 November 2023
Revised: 9 December 2023
Accepted: 11 December 2023
Published: 18 December 2023

1. Introduction

The assessment of the condition of historic buildings must be carried out according to a specific schedule and in accordance with relevant regulations [1]. The choice of repair method and type of structure strengthening [2,3] may apply both to historic buildings [4–6]—which can be several hundred years old—and to buildings erected relatively recently but damaged as a result of the progressive degradation of structural elements, exceptional loads caused by fire or war, errors in construction work, or low quality of building materials [7–9]. Historic buildings are a source of knowledge about the building technology of former times and, therefore, as symbols of cultural heritage, must be legally protected. Therefore, it is necessary to monitor and check their technical condition and to protect and appropriately strengthen them [10]. The poor conditions of historic buildings or the high costs of their revitalization do not qualify them for decommissioning and possible demolition. Due to their cultural heritage, they must be carefully assessed in order to be approved for safe operation.

Traditional methods for identifying defects and assessing their impact on the character of structural work may turn out to be unreliable when the condition of facilities that have operated for several hundred years is checked [11,12]. Conducting destructive tests in historic buildings requires obtaining appropriate permissions so that the number of destructive tests is limited to the necessary minimum. However, NDT methods remain, requiring people with an appropriate level of experience to carry out this type of testing. NDT using modern devices allows for a much easier diagnosis. For example, georadar, sonic and radar tests [13], and ground-penetrating radar (GPR) [14] enable the detection of internal defects in structural elements. Terrestrial laser scanning (TLS) is also a useful tool for architectural investigations of the external and internal structures of buildings [15,16]. TLS enables not only the accurate reflection of architectural details of the façades of historic buildings [17] but also, using reflection intensity, allows for the analysis and determination of the material from which a given structural element is made [18].

Examples of testing procedures using NDT or semi-destructive techniques (SDTs) are popular in the literature [19–21], especially due to increasingly widespread access to devices enabling this type of analysis [22,23]. Numerous described analyses and examples of specific studies [24,25] also confirm the validity of using these methods in historic buildings. NDT used in the assessment of historic buildings concerns both masonry structures and wooden structural elements [24,26–28].

In addition to the research and assessment of the actual condition of historic buildings, issues relating to the impact of harmful external factors on the properties of materials from which the structure was made have also been analyzed [29,30]. A useful tool for assessing the technical condition of historic buildings is historic building information modeling (HBIM) [31,32]. Multidisciplinary methodologies are used to implement building information modeling (BIM) for the analysis of historic buildings [3,5]. Taking into account a BIM flowchart supports the process of assessing the condition of the structure, as confirmed in [33,34]. Aspects taking into account sustainable development are also important during the process of condition assessment and renovation [35].

All the above-mentioned methods and tools used in the analysis of technical structural conditions enable the undertaking of a rational decision regarding potential repairs and strengthening [2,3]. Attention and care to the preservation of the original architecture, historical form, and detail requires a comprehensive approach in this scope [36]. The essence is the awareness that any interference in a material structure is intended to extend the time of safe operation of the building structure. This is why structural testing after the strengthening and repair stage plays such an important role [37]. When designing new structures and assessing the condition of structures that have been in use for several decades, it is important to meet technical regulations and the required level of safety [38–41]. In historic structures, even after strengthening, it is often not certain that safety reserves are at the required level. Therefore, more useful tools are being developed to support the process of solving problems relating to design, planning, management, erection, and operation processes. New solutions and ideas that solve technical and technological problems, which may appear in every single stage of projects, must be substantiated. The methodology of TRIZ (theory of inventive problem solving) may be a useful tool in this aspect [42]. This also applies to replacing existing materials while taking into account an effective eco-design strategy [43].

This paper presents an important problem, namely, changes in the load transfer and the redistribution of internal forces in 18th-century building structures. Examples of defects, damage, and human interference observed in historic buildings, affecting changes in the static schemes of structural elements, are described. A significant number of analyzed defects were observed in a historic pavilion building of a historic palace complex located in the northeastern part of Poland. Using this building structure as an example, not only were the existing defects described but numerical analyses concerning the impact of defects on static scheme changes of wooden roof truss were also carried out.

One of the problems discussed in this paper is the analysis of the wooden roof truss structure. The analysis and assessment of historic wooden structures were described in the literature regarding similar issues. Examples that describe the methodology for assessing the condition of historic wooden structures are [44,45], where in both cases, NDT methods were used. Correlations in the NDT tests, visual grading, and destructive tests were undertaken in [46]. In the work [47], the relationship between the results obtained from NDT and destructive tests was described by emphasizing how important it is to identify wood species. In this paper, the use of NDT methods to determine the material parameters of wooden elements with the highest possible accuracy is described. Moreover, the analysis using numerical calculations is presented. This article describes an attempt to determine the factors influencing the changes in the static scheme of the structures. This is what distinguishes analyses presented in this paper from previous contributions.

2. Defects Resulting in Static Scheme Changes in Structure

An important aspect of assessing a building's condition is to identify defects and determine the causes of negative destruction processes. In addition to the natural aging processes of structures, phenomena that accelerate the degradation of the facility often occur. However, it should be remembered that the main reason for the deterioration of the technical condition of buildings is the lack of ongoing repairs and proper facility maintenance by the owners. Historic buildings are under official conservation protection, which effectively prevents this. There are a number of destructive methods that allow for the determination of current material parameters. In historic buildings, carrying out such tests requires special permits. Therefore, whenever possible, NDT methods are used. An additional issue that may arise during the research and analysis of a structure's condition is the possibility of change in the static scheme. There can be several reasons for changes in static schemes. The most important are degradation of construction materials, excessive settlement of supports, excessive looseness in joints, and changes in the geometry of structural elements (change in dimensions, deflections due to changes in humidity, shrinkage, and rheological phenomena). Geometric changes in structural elements may occur in the elastic and inelastic range. Geometric changes in the inelastic range may be caused by rheological effects, brittle fracture or plasticization, biological and chemical corrosion, physical processes such as cyclic freezing and thawing, excessive moisture caused by leaky roofing or faulty ventilation [48,49], and washing out or dissolving small particles from the material structure. Presented below are examples of historic buildings from the 18th century in which, as a result of defects, changes in the static scheme of structural systems occurred.

2.1. Realization of Installation below the Foundation Level

An investigation of the destruction state was carried out on a historic three-story building dating from the second half of the 18th century, which was constructed as a traditional brick wall structure. In the building, there was no basement, and the foundation walls were made partly of cobblestone and full ceramic brick with lime mortar.

Installing pipes, for example, during the modernization of the sewerage network, close to buildings and below the level of foundations is associated with high risk. Deep excavation below the foundation level, without appropriate protection, results in changes in the natural structure of the soil. It may cause uneven settlement of the foundation and, consequently, cracking of load-bearing walls. In extreme cases, the installation pipe leaks, and the flowing water or sewage washes out small particles of soil under the foundations. This also leads to uncontrolled settlement in the foundations and cracking of the walls.

In the building presented in Figure 1, the width of the existing cracks significantly increased and new cracks appeared just after the completion of earthworks related to the modernization of the installation connection to the building. An inventory of the wall cracks is presented in Figure 1. As a result of the cracks, the method of load transfer in the external longitudinal wall was changed. The wall was divided into smaller fragments, which began to function as separate independent elements, similar to pillars, arches, and vaults.

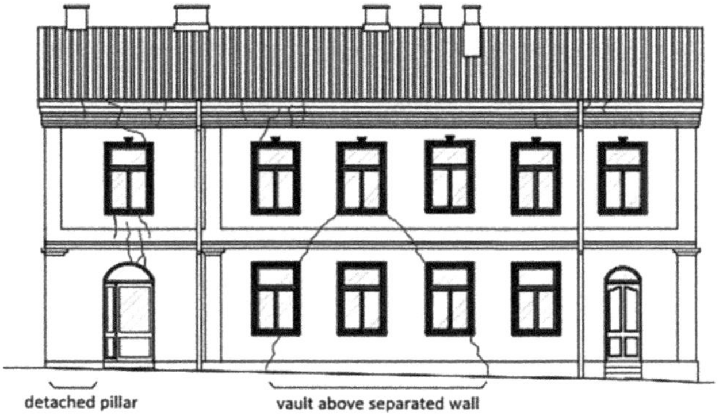

Figure 1. Separation of the wall fragments, due to scratches and cracks, into smaller parts that work independently.

2.2. Eccentrically Constructed Walls and Collapsed Vaults

Defects that cause changes in the static schemes were observed in a two-story building in which the basement walls were built at the turn of the 18th and 19th centuries. The above-ground walls were constructed in the second half of the 19th century, preserving the basement of the building that previously existed in this place. The arrangement of the basement walls does not correspond to the arrangement of the ground floor walls. The internal longitudinal load-bearing wall on the ground floor does not coincide with the basement wall and is based on the brick vault not on the wall located in the basement, as shown in Figure 2. The offset of the walls is approximately 90 cm.

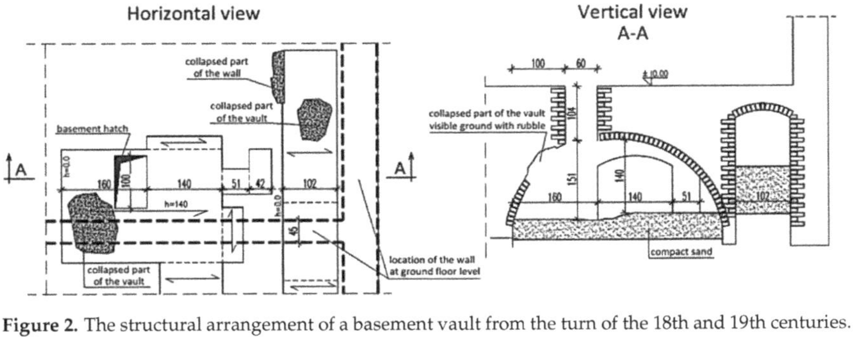

Figure 2. The structural arrangement of a basement vault from the turn of the 18th and 19th centuries.

The basement is partially filled with soil and rubble. In the place where the locally collapsed vault is located, part of the load-bearing wall on the ground floor has no support. Therefore, a natural secondary vault was created in the wall over the damaged vault. The load transferring was changed, and the forces were transmitted through wall fragments with appropriate support on the brick vault.

2.3. Destructive Impact of Rainwater

Leaking gutters and downpipes cause improper drainage of water from the roof, which results in dampness and subsequent degradation of the walls. In the long term, it may cause serious damage to the walls and foundations. Moisture in the walls is particularly dangerous in temperate climates in the winter when the temperature outside is around zero. Water freezes up and thaws out repeatedly in the wall cracks and pores of ceramic elements.

The result is visible in the form of plaster separations and the defects propagations, Figure 3. Bricks and mortar are damaged. The thickness and the load-bearing capacity of the wall is reduced. Vertical forces from higher floors are displaced in relation to the axis of the foundation walls, which additionally influences their uneven settlement.

Figure 3. Damage to a plaster and brick wall due to leaking drainpipes.

Draining rainwater into the ground close to the foundations, especially shallow foundations in non-cohesive soil, leads to the washing out of small particles from the soil framework. As a result, the soil structure is weakened, and the foundations settle. This is a common cause of cracks and scratches on the walls.

2.4. Consequences of Wall and Ceiling Reconstruction

In a three-story residential building from the end of the 18th century, numerous reconstructions were carried out during the building's operation. The originally homogeneous brick walls were transformed into separate parts with different technical conditions and strengths. Moreover, in the same building, during the reconstruction, wooden ceilings were demolished and replaced with reinforced concrete ceilings on steel beams. Ceiling slabs only provide lateral support for walls if they have adequate stiffness. The demolition of ceilings on several levels at the same time, leaving only wooden ceiling beams without sheathing, led to a change in the stiffness of the lateral support of the walls. When the sheathing was dismantled, the ceilings lost their stiffness and stopped providing continuous support for the walls. As a consequence, until new ceilings were constructed, there was a change in the method of lateral support and buckling length of the walls.

2.5. Construction of a Ceiling with Low-Stiffness Beams

An interesting case is a single-story production building from the end of the 18th century, which is covered with a gable roof supported with longitudinal walls and a wooden beam ceiling. During modernization work in the second half of the 20th century, thermal insulation was placed on the ceiling and a 7 cm thick floor of cement screed was made on it. The additional load on the ceiling increased the deflection in the ceiling beams, as shown in Figure 4a. Therefore, the ceiling structure required temporary support, as shown in Figure 4b. During repair work in 2022, all floor layers were removed from the ceiling. It did not completely eliminate the deflection in the ceiling beams. The beams remained permanently deflected due to rheological effects.

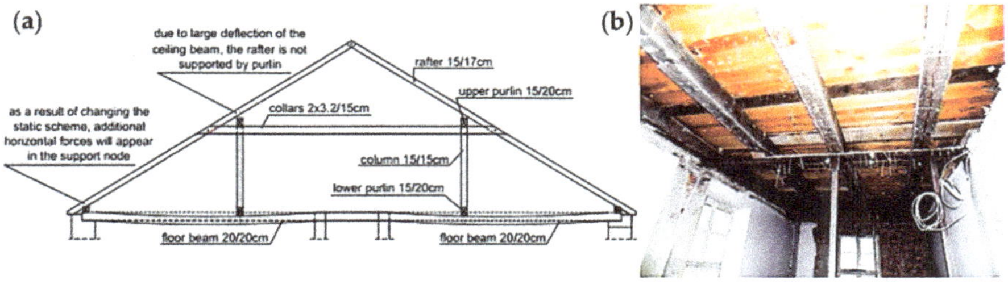

Figure 4. Deformation of a ceiling with wooden beams: (**a**) roof truss structure supported on the deformed ceiling and (**b**) view of the ceiling temporarily supported in order to limit the deflection.

The construction of ceilings using beams with insufficient stiffness leads to excessive deflections. The problem starts when other structural elements, e.g., a roof truss, rest on excessively deflected ceiling beams. The excessive deflection in the beams changes the support and distribution of internal forces in the structure based on such a ceiling. This leads to changes in the static scheme.

3. Analysis of the Destructive Stage of a Pavilion Building

The accumulation of defects and damages, which contribute to static scheme changes, can be presented in the example of an 18th-century pavilion building that is a part of the historic palace complex, as shown in Figure 5. During the assessment of the facility condition, the effects of significant degradation of individual structural elements were observed. The degradation of roof truss elements and the loss of proper support contributed to a change in the character of the work of the roof truss structure. In the analyzed case, as a result of the tests performed, the degree of collapse risk of the roof structure was assessed, which was particularly important for the planned renovation work.

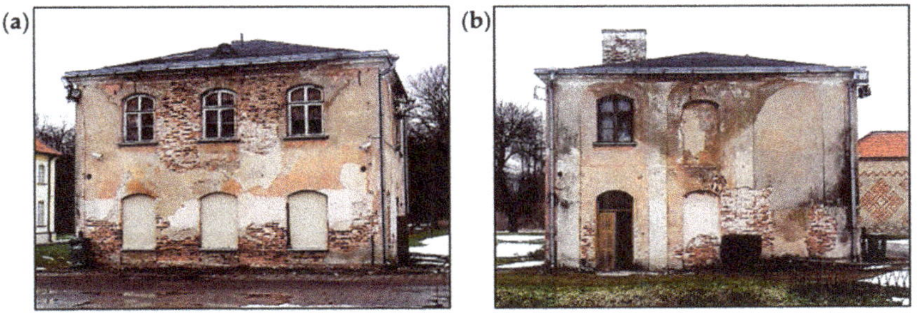

Figure 5. Pavilion building: (**a**) front elevation and (**b**) side elevation.

This section describes a number of defects found in the structural elements of the pavilion building. The description of defects is preceded by a presentation of the history and the structure of the building.

3.1. History of the Pavilion Building

The foundations of the pavilion building were laid in 1753. The construction and finishing work took several years. After 1840, the pavilion was intended to serve as a residence for factory officials. In 1915, during World War I, the palace buildings were partially blown up, and only the one-story walls of the pavilion building. remained In 1937, the palace buildings were rebuilt and used as a hospital. During World War II, the hospital was closed down, and the Soviet army was quartered in palace complex buildings. After 1941, a camp for several thousand prisoners of war was organized. After the end of the war, the hospital was reactivated, and the pavilion building was used for the needs of the hospital farm until 2003. There were offices on the ground floor, and the first floor was residential. Currently, the pavilion building is not in use. During the last 20 years, the condition of structural elements has been progressively degrading. The current arrangement of the main structural elements is described in the next section.

3.2. Building Structure

The pavilion building was constructed as a two-story building, without a basement, with an attic and covered with a hipped roof made as a wooden structure with traditional technology, as shown in Figure 5. The walls were made of solid ceramic bricks. The ceiling above the ground floor was made of reinforced concrete slabs, and the ceiling above the first floor was made entirely of wood with wooden beams. The roof truss was made of wood in a rafter–purlin structure. The ceilings and roof truss were built in 1937.

The foundation of the pavilion building was made of cobblestone with lime mortar. The lintels were made of arched brick. Some of the lintels, built during the reconstruction in 1937, were made of steel beams. The building was rebuilt several times. The effects of numerous brickworks and changes in the location of existing window and door openings were identified on the external walls.

3.3. Building Defects

The process of building condition assessment began with the foundations, which were made of cobblestone arranged in a masonry bond manner. Based on the excavations, it was found that the foundation stone wall has a compact structure. There was no mortar in the joints on the upper surface of the foundations, but lime mortar in a satisfactory condition was found deeper in the joints. The foundations were wider than the foundation walls by approximately 40 cm, and the building corners by approximately 60 cm, as shown in Figure 6.

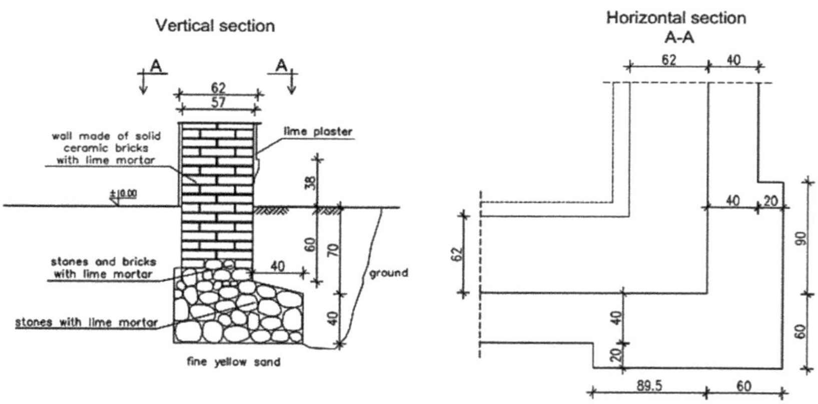

Figure 6. Vertical and horizontal sections of the foundation extension in the building corner.

The foundation walls had the same width as the above-ground walls. Their bottom was approximately 60 cm below ground level. The upper part of the foundation walls was made of ceramic bricks with lime mortar, and the lower one was made of cobblestone and solid ceramic bricks with lime mortar. The mortar on the wall surface was degraded, and its strength was close to 0 MPa. Small scratches, which were caused by uneven settlement in the building, were found on the foundation walls just above ground level.

Defects in the walls aboveground were concentrated in the corners of the building, as shown in Figure 7. The bricks and lime mortar in the surface layers were significantly degraded in many places and had a strength close to 0 MPa. In the deeper parts of the wall, the mortar had a color typical of lime mortar, and its strength was estimated at approximately 0.2 MPa. The strength of bricks determined using destructive tests in the laboratory met the standard criteria.

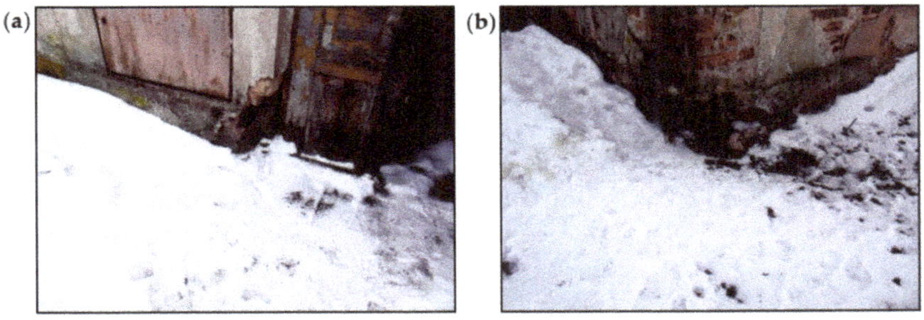

Figure 7. View of defects in external brick walls: (**a**) near a door opening and (**b**) in the corner of the building.

The significant degradations of the brick walls and plaster losses visible on the surface of the building's façade were observed. Above the windows in the arched lintels, damage was indicated in the form of scratches in the plaster and the loss of mortar, as shown in Figures 8 and 9. The causes of damage to the brick arches above the windows and doors were uneven settlement of the foundations, deformations of the walls, dampness of the walls, and freezing and thawing of water inside the wall structure.

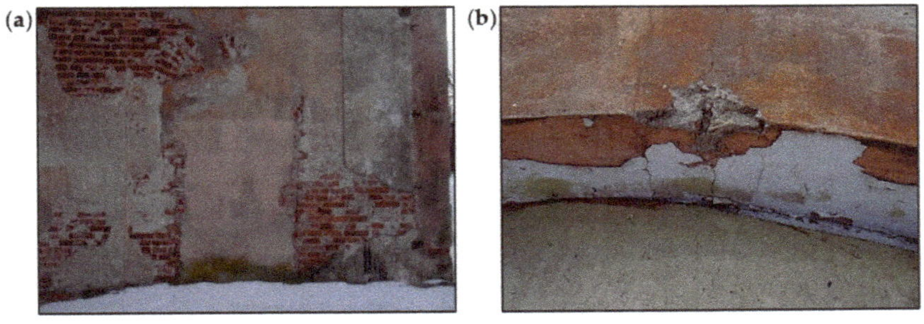

Figure 8. Damage to external walls: (**a**) plaster losses and bricked-up door opening and (**b**) crack in the arch above the window.

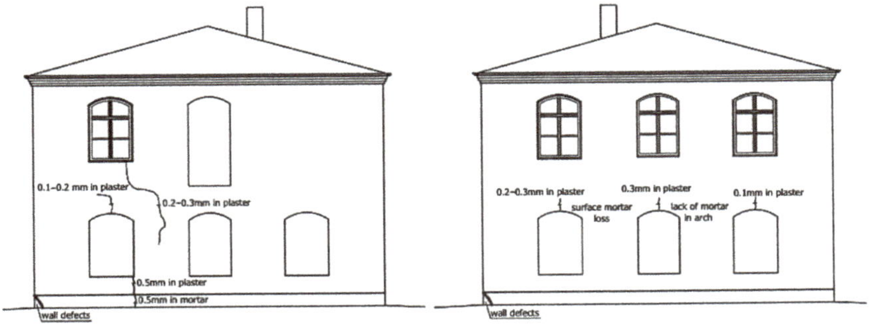

Figure 9. Inventory of scratches on two external walls.

Large areas of the walls throughout the building were damp. In selected places, wall humidity measurements were performed using the electro-resistance method. The greatest dampness occurred at the ground level and just above it.

Water is a significant destructive factor in the case of brick walls. Capillary properties of the brick cause water to come up from the ground to the height of several meters of the wall, simultaneously transporting water-soluble salts. Salt crystallization in the surface layers of the bricks causes the mechanical destruction of clay material structure, resulting in crumbling and brick fragments falling off the walls. Salts, due to the hydrolysis process, change the reaction of water to slightly acidic or alkaline. This results in the destruction of mortar and the dissolution of clay material in bricks.

In the pavilion building, the effects of capillary phenomena were visible on the walls, especially on the lower parts. Low temperatures and water freezing led to mechanical corrosion of the walls. Locally, the corrosion reached a depth of several centimeters inside the wall.

In the walls with lower dampness, low temperatures also cause systematic separation of plaster coatings after subsequent winter periods. After the wall is exposed (the plaster is chipped off or it falls off on its own), the effects of brick corrosion are visible in the form of brick fragments that flake, crumble, and falling off.

There were also visible effects of the corrosion caused by rainwater, flowing down from the top of the walls. The capillary properties of bricks and mortars cause rainwater to be excessively absorbed in the places where the walls are damaged. In such places, due to the difficult evaporation of water, significant damage occurs when water freezes in the winter.

Another factor causing corrosion is the active growth of microorganisms in walls with increased humidity. As a result of algae growth, substrate degradation occurs due to the secretion of organic acids. Acidic products of algae metabolism change the chemical composition of the water in wall capillaries, which additionally increases corrosion.

Based on the analysis of external walls, the following defects and findings were noted: significant surface damage to the walls, local defects just above ground level, local scratches, relatively low compressive strength of the bricks (class 5), and relatively low compressive strength of the lime mortar (approximately 0.2 MPa). However, no significant vertical or horizontal deformations of the external walls were detected. There were also no cracks propagating through the entire width of the wall. Due to the above-mentioned factors, it should be stated that the technical condition of the external walls was sufficient.

The ceiling above the ground floor was made of reinforced concrete slabs and ribs, as shown in Figure 10a. The ceiling was covered from below with a suspended ceiling made of cement-lime plaster on steel mesh. Between the ceiling and the ceiling ribs, there was a soffit made of boards. Two opencasts of the ceiling were made: from the bottom and the top (Figure 11a). Based on the opencasts, it was found that the ceiling concrete slab was made with a thickness of 7 cm, and the main reinforcement of the slab consisted of smooth bars with a diameter of 8 cm every 17 cm. There was no distribution reinforcement in the slab on a section of 40 cm. The ribs were 11 cm × 25 cm (height measured including the ceiling slab). In the ribs, no reinforcement was found in the upper zone, but just under the reinforced concrete slab, there were two bars with a diameter of 16 cm. In the ribs, no stirrups were found on a section of 30 cm.

Sclerometric tests showed that the ceiling was made of C8/10 class concrete. Due to the lack of stirrups in the ribs in the support zone, it should be considered that the shear load capacity of the reinforced concrete ribs was insufficient. Due to the lack of stirrups or their too large spacing in the ribs, the lack of distribution reinforcement in the ceiling slab, and the low-quality concrete, it should be considered that the condition of the reinforced concrete ceiling was poor.

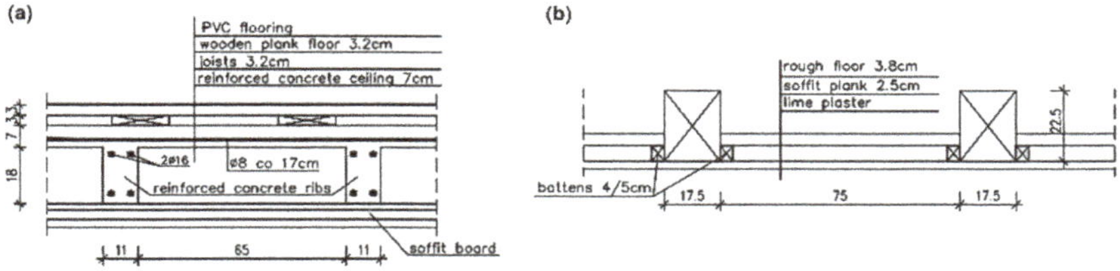

Figure 10. Ceiling sections: (**a**) concrete ceiling above the ground floor and (**b**) wooden ceiling above the first floor.

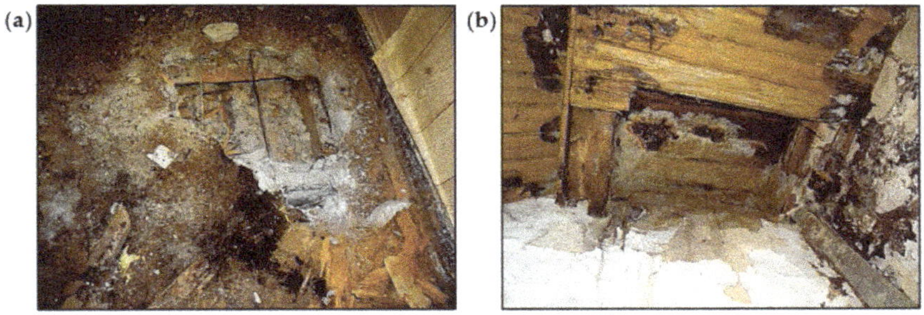

Figure 11. Ceiling opencasts: (**a**) from the top of the reinforcement concrete ceiling and (**b**) from the bottom of the wooden ceiling.

The ceiling above the first floor was made of wooden beams, as shown in Figure 10b. Leaks in the roof covering caused dampness of the ceiling and contributed to the development of biological corrosion. The ceiling above the staircase and the room on the southeast side completely collapsed, as shown in Figure 12. The remaining parts of the ceiling were very damp, as shown in Figure 11b. There were plaster defects and cavities in the sections close to the collapsed parts of the ceiling. The ceiling beams were significantly damaged as a result of biological corrosion. The remaining parts of the ceiling were in danger of collapse. The overall condition of the wooden ceiling was poor.

Leaks in the roof covering caused dampness in the wooden elements of the roof truss, and the biological corrosion developed. Wooden elements close to the chimney eaves and a leaky roof covering were significantly damaged. Some of the rafters were damaged in the places where they rested on the roofing wall plates. The collapse of the wooden ceiling above the first floor caused the pillar walls of the roof truss to be supported only on the brick walls and fragments of the remaining ceiling. The pillar walls above the staircase and the room on the southeast side had no support.

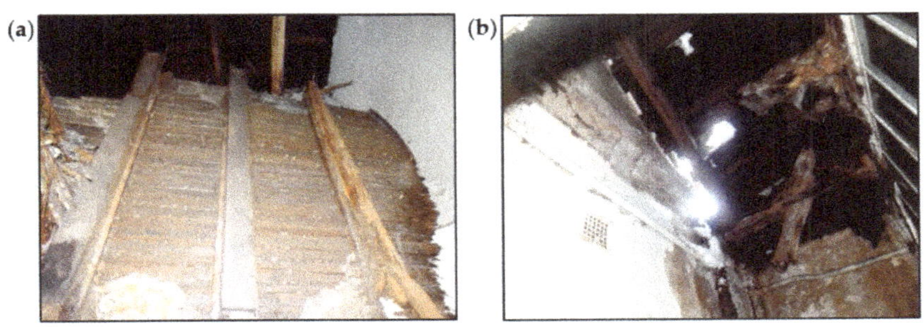

Figure 12. *Cont.*

(c) (d)

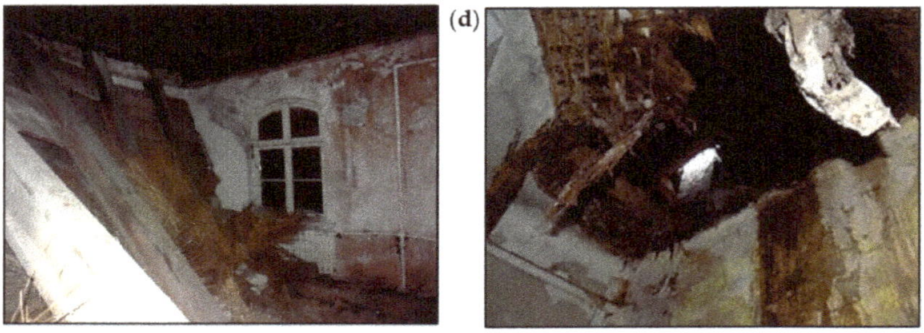

Figure 12. View of the collapsed wooden ceiling: (**a**,**b**) above the staircase and (**c**,**d**) above a room on the first floor.

4. Tests of Materials and Measurements Carried out during the Condition Assessment of the Pavilion Building

4.1. Destructive Testing of Bricks

The quality of bricks in the walls was assessed based on laboratory tests of compressive strength. Drilled holes (Figure 13a) were made in the walls to collect samples for laboratory tests. During drilling, the mortar was washed out, which indicated that its compressive strength was equal to 0.1–0.2 MPa according to [50]. Strength tests of bricks (Figure 13b) were performed in accordance with applicable standards [51]. For three samples that were taken for testing, a total of 10 tests were performed. The average compressive strength of the individual brick samples was equal to 6.6 MPa, 7.9 MPa, and 4.7 MPa, as presented in Table 1. Based on the obtained strengths, the tested bricks should be classified as class 5, which is the lowest class of bricks. The strength of the lime mortar in the walls should be estimated at 0.2 MPa.

Table 1. Compressive strength of brick samples obtained from laboratory tests.

Sample Number	Core Diameter (mm)	Maximum Compressive Force (kN)	Strength (MPa)	Average Value of Compressive Strength (MPa)
1.1	54	9.9	4.3	6.6
1.2	54	18.9	8.2	
1.3	54	11.0	4.8	
1.4	54	20.6	9.0	
2.1	54	20.8	9.1	7.9
2.2	54	11.6	5.1	
2.3	54	21.6	9.4	
3.1	53	10.9	4.9	4.7
3.2	54	12.3	5.4	
3.3	53	8.4	3.8	

According to the findings obtained from destructive tests and taking into consideration the visual evaluation of the defects, the technical condition of the walls was classified as sufficient.

4.2. Wood Testing Using a Woodtester

During the condition assessment of the wooden roof truss, NDT methods were used such as measuring the wood density using a Woodtester and measuring the wood humidity using a moisture meter. The most important reasons for the poor condition of the roof truss were leaks in the roof covering, provisional repairs of defects, lack of proper flashing of the chimneys and gutters, lack of periodic inspections, and supplementary impregnation of the wooden truss structure with fungicides and insecticides. Humidity tests showed that the moisture content of the wooden elements of the roof truss was 30–50%.

(a) (b)

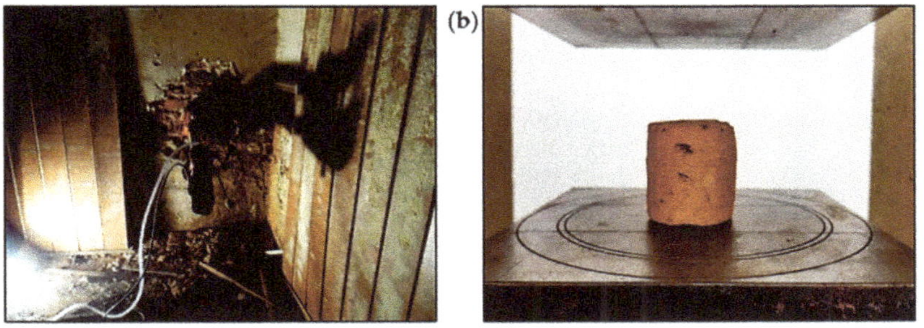

Figure 13. Destructive testing: (**a**) cutting out brick samples from the wall and (**b**) the strength test of a brick.

The condition of the roof truss structure was thoroughly assessed visually. Progress of the biological corrosion and damage to the elements were very significant.

Despite the difficult access to the roof truss elements, due to the ceiling collapse, partial tests were performed using a Woodtester, as shown in Figure 14. Based on the depth of needle penetration into the wood, it was possible to estimate the density of the wood and then the average modulus of elasticity. In the case of the analyzed roof truss, the degradation of the cross-sections of the elements was significant. Based on the Woodtester tests, the average modulus of elasticity of the wood was estimated and applied in numerical calculations.

(a) (b)

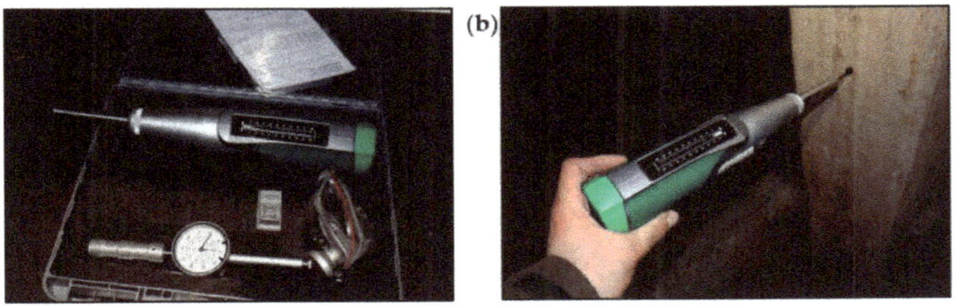

Figure 14. Woodtester: (**a**) view of the device and accessories and (**b**) during the test.

4.3. Terrestrial Laser Scanning of the Pavilion Building's External Walls

Another NDT method used during the assessment of the pavilion building condition was TLS, which allows for performing comprehensive measurements of a building's geometry. As a result, a point cloud is obtained, where each point has four coordinates assigned to it. Three of them, X, Y, and Z, are shown in the local coordinate system of the scanner and can be transposed to any geodetic system. The fourth coordinate describes the reflection intensity of the laser beam from the measured element. Scanning devices can collect data from a maximum distance of 300 m, with a horizontal field of view up to 360 degrees and a vertical plane up to 270 degrees. Measurements are performed with an accuracy of up to 1 mm.

TLS technology allows for performing analyses in the field of inventory of the actual shape and current condition of building structures as well as measurements of displacements and deformations in structural elements and buildings. A laser scanner is a useful tool for collecting geometrical data about the technical condition of building objects. Based on the measurements, it provides data on the shape and dimensions of the analyzed object. At the same time, defects in the form of deformations, cracks, or other surface damage are inventoried. The data obtained during the measurements are processed using computer programs, which allows for obtaining a virtual, spatial copy of a measured building [15–18]. Additionally, thanks to digital photos taken during the measurement, a three-dimensional model can be viewed in natural colors.

The 3D laser scanning and specialized computer programs allow for more complete, more accurate, and more legible documentation of the building. The obtained information is very detailed, so the scope of its application is wide. For this reason, scanners are increasingly replacing other measuring tools.

In the case of the pavilion building, laser scanning was performed. External walls were scanned, and the digital point cloud was obtained. Views of the pavilion building after the appropriate processing and development of the point cloud are shown in Figure 15. Thanks to precise measurements, the actual geometry of the pavilion building was available, as shown in Figure 15a. Significant advantages include the possibility to observe in detail the defects in external brick walls and also to determine the depth of these defects, as shown in Figure 15d.

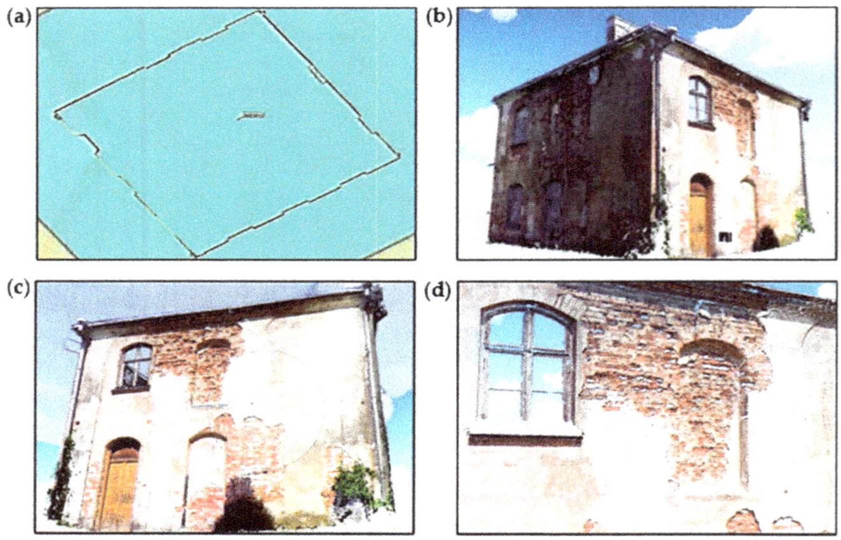

Figure 15. Images obtained based on the point cloud: (a) horizontal projection with building dimensions; (b,c) a view of the external walls and (d) a view of defects in the brick wall.

5. Numerical Analysis of the Wooden Roof Structure of the Pavilion Building

The intensive degradation process of the wooden roof structure began with damage to the roof covering. It directly caused considerable dampness of the upper parts of the external walls, the roof truss, and the wooden ceiling above the first floor on which the roof truss was supported. The wooden roof truss and ceiling were attacked by fungi and were seriously damaged. The rapid progress of biological corrosion weakened and damaged the support zones of the ceiling beams. This resulted in the collapse of the ceiling parts (beams 1–5) supporting the roof truss, as shown in Figure 16. The roof truss structure partially lost its support. Despite serious damage, the roof truss did not collapse and remained in a state of unstable balance.

A spatial computational model of the roof truss structure was prepared in order to precisely assess the degree of effort of structural elements. Based on the Woodtester results, taking into account the actual material properties and rheology of the wooden elements, the average final modulus of wood elasticity was estimated as $E_{mean,fin}$ = 3667 MPa [41].

The flexibility of the roofing wall plate support in the horizontal direction of 500 kN/m was assumed for all calculation variants. The loads were determined according to [38–40].

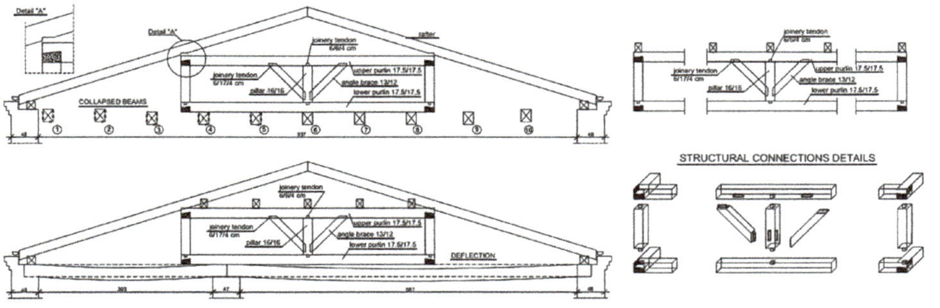

Figure 16. View of the roof truss structure supported by a partially destroyed ceiling.

In the numerical analysis, computational models of the roof truss structure were made taking into account:

- The low value of the elastic modulus of the wood;
- The influence of humidity and rheology;
- The nature of the load and duration of the load (DOL) effect;
- The flexibility of supports in the horizontal direction (support on roofing wall plates);
- The degradation of some structural elements;
- The lack of supports due to the collapse of the wooden ceiling;
- The impact of deformation in the main structural elements on the secondary elements based on them;
- The character of the joints' work (some joints can only transfer compressive forces);
- The discontinuities (looseness) in the joints.

In the first case (variant I), the calculations showed that the ceiling beam not loaded with the pillar wall had a greater deflection value than the adjacent ceiling beam loaded with the pillar wall. The displacement diagram showed that the pillar wall was not supported on the ceiling beam, but the ceiling beam was suspended to the pillar wall. Observation of this phenomenon was possible for permanent loads acting on the structure, as shown in Figure 17. Since the loads transferred from the roof were relatively small, tensile forces occurred in the connection of the ceiling beam with the pillar wall, which was inconsistent with reality.

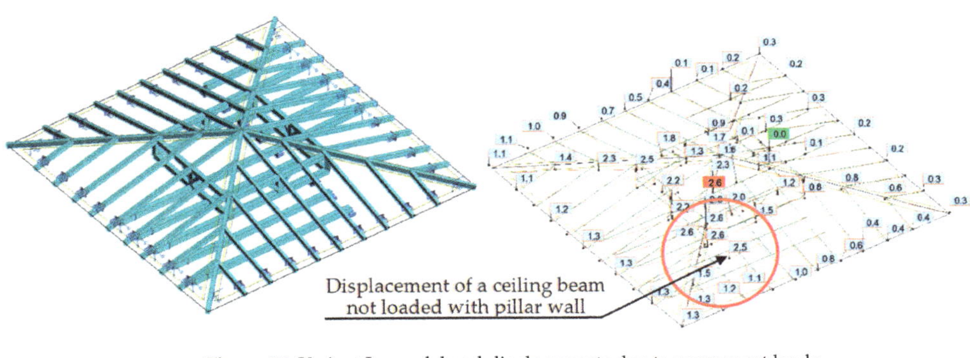

Figure 17. Variant I—model and displacements due to permanent loads.

In the second case (variant II), it was assumed that the pillars and swords could not transfer tensile forces due to the character of the work of the carpentry joints. For the same permanent loads, the deflection in the ceiling beam supporting the pillar wall was greater than the deflection in the adjacent ceiling beam not loaded with the pillar wall, as shown in Figure 18.

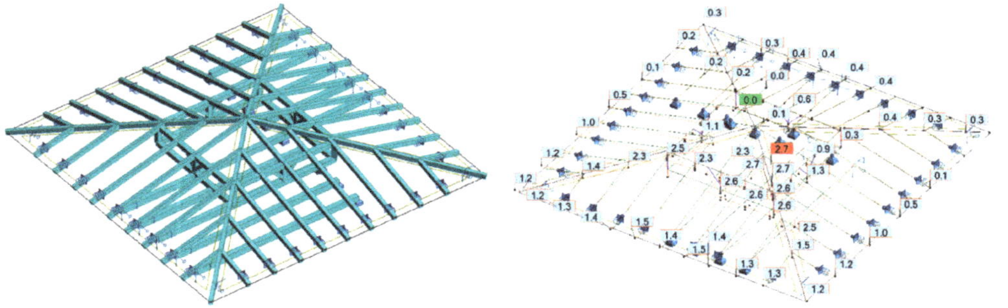

Figure 18. Variant II—model and displacements due to permanent loads; the pillars and swords can transfer only compressive forces.

In the third case (variant III), the same permanent loads and the same assumption that the pillars and swords can transfer only compressive forces were applied. The modification assumed that the supports of the roof truss in the form of the remaining ceiling beams were replaced with elastic supports with flexibility corresponding to the ceiling beams. Moreover, the displacements in the support points were assumed to be equal to 2.5 cm to correspond to the deflection in the ceiling beam under the ceiling's own weight, as shown in Figure 19.

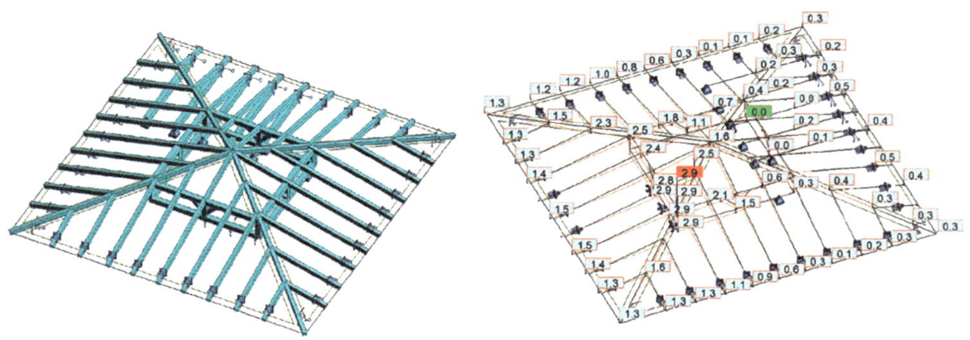

Figure 19. Variant III—model and displacements due to permanent loads; the pillars and swords can transfer only compressive forces; and flexible supports with the initial deflection of the ceiling beams.

In the fourth case (variant IV), calculations were made using the same assumptions as in variant III. In addition to the permanent loads, the loads from snow were also taken into account, as shown in Figure 20.

In the presented diagrams, different values for displacements in the pillar wall were obtained. It mainly depended on the method of support and the possibility of transferring tensile forces through the pillars and swords. Based on the analyses and calculations performed, it was confirmed that the spatial computational model of the roof truss is susceptible to many factors, which have a significant influence on the displacement results.

If the stiffness of the supports (in this case the ceiling beams) was low, the first calculations showed that the roof truss was not supported by the ceiling beams, but the ceiling beams were suspended to the roof truss, which is inconsistent with reality. It was necessary to analyze the character of the work of the carpentry joints. On this basis, an assumption was made that some elements of the roof truss were unable to transfer tensile forces. These were the elements that were connected with the other elements with carpentry joints. After taking this assumption into account in the calculations, the internal forces and displacements in the structure changed significantly.

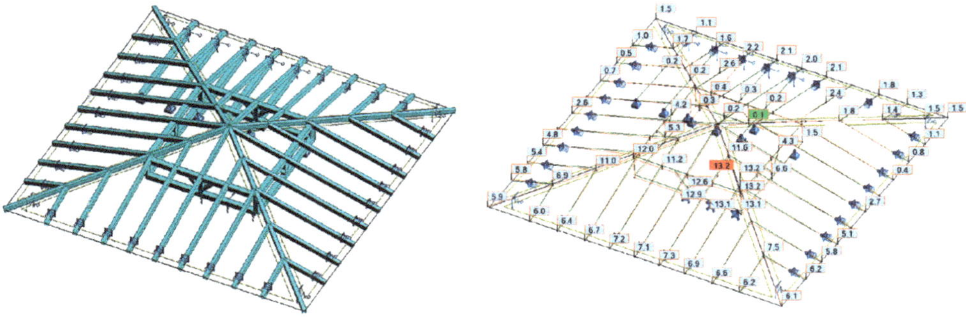

Figure 20. Variant IV—model and displacements due to permanent loads and snow, pillars, and swords can transfer only compressive forces and flexible supports with the initial deflection of the ceiling beams.

6. Discussion

All defects described in this article have a negative impact on the technical condition of structural elements. The destructive impact of external factors can be divided into two groups:

- Group 1—destructive processes cause a gradual deterioration of the strength parameters of structural elements until their load-bearing capacity is reached.
- Group 2—destructive processes, at some stage, cause a change in the static scheme, as a result of load-bearing capacity loss of a single or several elements or as a result of excessive deformations in individual structural elements.

In the brick walls, making additional openings or filling existing openings with non-structural materials causes several fragments of walls may be not connected to each other and to work independently. As a result of uneven settlement of the foundations, cracks in the walls appear. Natural dilatations are created and individual wall fragments work independently of each other. Above the separated fragments, the wall can maintain its continuity, e.g., by creating a spontaneous secondary arch.

In the case of wooden structures with carpentry joints, the distribution of forces and stresses is more difficult to interpret. As a result of the redistribution of internal forces and change in the static scheme, the structure may collapse or maintain spatial stiffness and still continue the safe transfer of loads in the new static scheme.

During the condition assessment process of historic buildings, it is necessary to conduct a detailed analysis, taking into account as many of the previously mentioned factors as possible. The analysis of masonry structures should consider, among others, wall re-building, the variety of materials from which parts of the wall were made, ground conditions, etc. In the case of spatial roof truss structures, as a result of progressive degradation, damage propagation, and reduced stiffness of structural elements, a change in the static scheme may occur. As a result, tensile forces appear in elements that previously had compressive forces. If the carpentry joints are unable to transfer tensile forces, a secondary change in the static scheme occurs. For the computational calculations, it is necessary to assess whether the displacements in the structure are consistent with the actual situation. It should also be checked whether the obtained internal forces can actually occur. Due to the character of the work of the supports and joints, elements have a limited ability to transfer various types of forces. During the computational analysis, carpentry joints in traditional wooden structures should be carefully examined because, depending on their type, they can transfer only compressive forces and cannot transfer tensile forces.

Multithreaded analyses are always challenging, despite the considerable experience of those who perform them. One of the most important repercussions, resulting from the conducted research, is a statement that the identification of the appropriate static scheme is often a difficult and demanding task. Errors may appear in the form of oversimplifications of computational models and material parameters. Appropriate methodologies and tools should be used to support the assessment process, especially in order to notice changes in the static schemes of the structure. The main implication of this study is the confirmation of the validity of using NDT methods in the condition assessment process.

7. Conclusions

Research and analyses of the structures presented in this article lead to the conclusion that many processes and actions related to progressive wear, changes in use, repairs, and reconstructions lead to changes in static schemes. Some of them may have a local impact, but often, the changes affect the static work of almost the entire structure. Obvious situations are removing or changing the flexibility of the supports and changing the stiffness of structural members and connections.

The decrease in the strength properties of building materials is caused by:

- Corrosion and anthropogenic processes;
- The aging of materials;
- Changes in loads acting on the structure;
- Changes in environmental conditions (temperature, humidity).

These factors cause the loss of load-bearing capacity of individual structural elements. In extreme cases, it results in failure, or the structure remains stable, but redistribution of internal forces and changes in the static scheme occur.

Taking into consideration aspects such as the impact of rheological phenomena, the effects of DOL, humidity, temperature, the actual technical parameters of materials, including degradation, and the character of joints is important in the process of assessing the condition of degraded historic building structures. All defects in structures contribute to changes in the static schemes. Based on the research presented in this article and the results obtained, the possibility of changes in the static schemes should be considered. This aspect requires emphasis because it has not been strongly promoted in the cases of the condition assessment of structures, which were previously described in the literature. The main limitation is the lack of dedicated guidelines for this purpose. However, research and the application of different methods are a part of the expected future development.

Author Contributions: Conceptualization, J.R.K., K.Z. and M.M.; methodology, K.Z.; software, K.Z.; validation, K.Z., M.M. and A.S.-S.; formal analysis, J.R.K.; investigation, K.Z.; resources, K.Z.; data curation, M.M. and A.S.-S.; writing—original draft preparation, J.R.K. and M.M.; writing—review and editing, M.M.; visualization, M.M. and A.S.-S.; supervision, J.R.K.; project administration, M.M.; funding acquisition, J.R.K. All authors have read and agreed to the published version of the manuscript.

Funding: This research was carried out within the scope of work no. WZ/WB-IIL/4/2023 and WI/WB-IIL/9/2022 at the Bialystok University of Technology and financed from the resources for science of the Ministry of Science and Higher Education of Poland.

Institutional Review Board Statement: Not applicable.

Informed Consent Statement: Not applicable.

Data Availability Statement: The data presented in this study are available on request from the corresponding author.

Conflicts of Interest: The authors declare no conflict of interest.

References

1. Riggio, M.; D'Ayala, D.; Parisi, M.A.; Tardini, C. Assessment of Heritage Timber Structures: Review of Standards, Guidelines and Procedures. *J. Cult. Herit.* **2018**, *31*, 220–235. [CrossRef]
2. Branco, J.M.; Descamps, T.; Tsakanika, E. *Repair and Strengthening of Traditional Timber Roof and Floor Structures BT—Strengthening and Retrofitting of Existing Structures*; Costa, A., Arêde, A., Varum, H., Eds.; Springer: Singapore, 2018; pp. 113–138, ISBN 978-981-10-5858-5.
3. Fofiu, M.; Marius, M.; Onescu, I.; Palade, M.; Olariu Gigi, O.G.; Popovici, T. Case Study of Consolidation, Rehabilitation, and Restoration of the "Sfantul Filimon" Monastery. In Proceedings of the COHESION 2021, Timisoara, Romania, 12–14 November 2021.
4. Mosoarca, M.; Gioncu, V. Failure Mechanisms for Historical Religious Buildings in Romanian Seismic Areas. *J. Cult. Herit.* **2013**, *14*, e65–e72. [CrossRef]
5. Lo Monaco, A.; Grillanda, N.; Onescu, I.; Fofiu, M.; Clementi, F.; D'Amato, M.; Formisano, A.; Milani, G.; Mosoarca, M. Seismic Assessment of Romanian Orthodox Masonry Churches in the Banat Area through a Multi-Level Analysis Framework. *Eng. Fail. Anal.* **2023**, *153*, 107539. [CrossRef]
6. Fazzi, E.; Galassi, S.; Misseri, G.; Rovero, L. Seismic Vulnerability Assessment of the Benedictine Basilica Typology in Central Italy. *J. Build. Eng.* **2021**, *43*, 102897. [CrossRef]
7. Krentowski, J.R. Assessment of Destructive Impact of Different Factors on Concrete Structures Durability. *Materials* **2022**, *15*, 225. [CrossRef]
8. Kosior-Kazberuk, M.; Krentowski, J.R.; Wardach, M. Diagnostics of the RC Roofing Structure of the 100-Year-Old Municipal Theatre Facility. *Materials* **2022**, *15*, 7438. [CrossRef]
9. Drobiec, Ł.; Grzyb, K.; Zając, J. Analysis of Reasons for the Structural Collapse of Historic Buildings. *Sustainability* **2021**, *13*, 10058. [CrossRef]
10. Krentowski, J.; Chyzy, T.; Dunaj, P. Sudden Collapse of a 19th-Century Masonry Structure during Its Renovation Process. *Eng. Fail. Anal.* **2017**, *82*, 540–553. [CrossRef]
11. Sánchez-Beitia, S.; Luengas-Carreño, D.; Crespo De Antonio, M. The Presence of Secondary Creep in Historic Masonry Constructions: A Hidden Problem. *Eng. Fail. Anal.* **2017**, *82*, 315–326. [CrossRef]
12. Sánchez-Beitia, S.; Schueremans, L. The Hole Drilling Technique for on Site Deduction of the Stresses States in Stone Masonry by Using Eight Strain Gages. *Constr. Build. Mater.* **2009**, *23*, 2041–2046. [CrossRef]
13. Binda, L.; Saisi, A.; Tiraboschi, C.; Valle, S.; Colla, C.; Forde, M. Application of Sonic and Radar Tests on the Piers and Walls of the Cathedral of Noto. *Constr. Build. Mater.* **2003**, *17*, 613–627. [CrossRef]
14. Lampropoulos, K.C.; Moropoulou, A.; Korres, M. Ground Penetrating Radar Prospection of the Construction Phases of the Holy Aedicula of the Holy Sepulchre in Correlation with Architectural Analysis. *Constr. Build. Mater.* **2017**, *155*, 307–322. [CrossRef]
15. Pawlowicz, J.A.; Szafranko, E. Application of Reverse Engineering in Modelling of Rural Buildings of Religious Worship. *Eng. For. Rural. Dev.* **2016**, *15*, 762–766.
16. Szafranko, E.; Pawlowicz, J. Inventory of Agricultural Building Objects Based on Data Obtained from Measurements by Laser Scanning. In Proceedings of the Engineering for Rural Development, Jelgava, Latvia, 20–22 May 2015.
17. Pawłowicz, J.A. Importance of Laser Scanning Resolution in the Process of Recreating the Architectural Details of Historical Buildings. *IOP Conf. Ser. Mater. Sci. Eng.* **2017**, *245*, 052038. [CrossRef]
18. Pawłowicz, J.A. Impact of Physical Properties of Different Materials on the Quality of Data Obtained by Means of 3d Laser Scanning. *Mater. Today Proc.* **2018**, *5*, 1997–2001. [CrossRef]
19. Binda, L.; Saisi, A.; Tiraboschi, C. Investigation Procedures for the Diagnosis of Historic Masonries. *Constr. Build. Mater.* **2000**, *14*, 199–233. [CrossRef]
20. Jaskowska-Lemańska, J.; Przesmycka, E. Semi-Destructive and Non-Destructive Tests of Timber Structure of Various Moisture Contents. *Materials* **2021**, *14*, 96. [CrossRef]
21. Hulimka, J.; Kałuża, M.; Kubica, J. Failure and Overhaul of a Historic Brick Tower. *Eng. Fail. Anal.* **2019**, *102*, 46–59. [CrossRef]
22. Schabowicz, K. Non-Destructive Testing of Materials in Civil Engineering. *Materials* **2019**, *12*, 3237. [CrossRef]
23. Wardach, M.; Krentowski, J.R.; Knyziak, P. Degradation Analyses of Systemic Large-Panel Buildings Using Comparative Testing during Demolition. *Materials* **2022**, *15*, 3770. [CrossRef]
24. Jiao, J.; Xia, Q.; Shi, F. Nondestructive Inspection of a Brick–Timber Structure in a Modern Architectural Heritage Building: Lecture Hall of the Anyuan Miners' Club, China. *Front. Archit. Res.* **2019**, *8*, 348–358. [CrossRef]
25. Briceño, C.; Noel, M.F.; Chácara, C.; Aguilar, R. Integration of Non-Destructive Testing, Numerical Simulations, and Simplified Analytical Tools for Assessing the Structural Performance of Historical Adobe Buildings. *Constr. Build. Mater.* **2021**, *290*, 123224. [CrossRef]
26. Nowak, T.P.; Jasieńko, J.; Hamrol-Bielecka, K. In Situ Assessment of Structural Timber Using the Resistance Drilling Method—Evaluation of Usefulness. *Constr. Build. Mater.* **2016**, *102*, 403–415. [CrossRef]
27. Bajno, D.; Bednarz, Ł.; Nowak, T. Problems Relating to Assessment, Repair and Restoration of Wooden Roof Structures in Historic Buildings, as Exemplified by Two Case Studies in Southern Poland. *Adv. Mater. Res.* **2013**, *778*, 888–894. [CrossRef]
28. Onescu, E.; Onescu, I.; Mosoarca, M. The Impact of Timber Roof Framework Over Historical Masonry Structures. *IOP Conf. Ser. Mater. Sci. Eng.* **2019**, *603*, 042030. [CrossRef]

29. Vitiello, V.; Castelluccio, R.; Del Rio Merino, M. Experimental Research to Evaluate the Percentage Change of Thermal and Mechanical Performances of Bricks in Historical Buildings Due to Moisture. *Constr. Build. Mater.* **2020**, *244*, 118107. [CrossRef]
30. Mosoarca, M.; Keller, A.I.; Petrus, C.; Racolta, A. Failure Analysis of Historical Buildings Due to Climate Change. *Eng. Fail. Anal.* **2017**, *82*, 666–680. [CrossRef]
31. Garcia-Gago, J.; Sánchez-Aparicio, L.J.; Soilán, M.; González-Aguilera, D. HBIM for Supporting the Diagnosis of Historical Buildings: Case Study of the Master Gate of San Francisco in Portugal. *Autom. Constr.* **2022**, *141*, 104453. [CrossRef]
32. Mora, R.; Sánchez-Aparicio, L.J.; Maté-González, M.Á.; García-Álvarez, J.; Sánchez-Aparicio, M.; González-Aguilera, D. An Historical Building Information Modelling Approach for the Preventive Conservation of Historical Constructions: Application to the Historical Library of Salamanca. *Autom. Constr.* **2021**, *121*, 103449. [CrossRef]
33. Massafra, A.; Prati, D.; Predari, G.; Gulli, R. Wooden Truss Analysis, Preservation Strategies, and Digital Documentation through Parametric 3D Modeling and HBIM Workflow. *Sustainability* **2020**, *12*, 4975. [CrossRef]
34. Barrile, V.; Bernardo, E.; Bilotta, G. An Experimental HBIM Processing: Innovative Tool for 3D Model Reconstruction of Morpho-Typological Phases for the Cultural Heritage. *Remote Sens.* **2022**, *14*, 1288. [CrossRef]
35. Bajno, D.; Grzybowska, A.; Bednarz, Ł. Old and Modern Wooden Buildings in the Context of Sustainable Development. *Energies* **2021**, *14*, 5975. [CrossRef]
36. Akcay, C.; Şolt, A.; Korkmaz, N.M.; Sayin, B. A Proposal for the Reconstruction of a Historical Masonry Building Constructed in Ottoman Era (Istanbul). *J. Build. Eng.* **2020**, *32*, 101493. [CrossRef]
37. Kwiecień, A.; Kuboń, P. Dynamic Analysis of Damaged Masonry Building Repaired with the Flexible Joint Method. *Arch. Civil. Eng.* **2012**, *58*, 39–55. [CrossRef]
38. *EN 1991-1-1:2004-Eurocode 1*; Actions on Structures-Part 1-1: General. Actions-Densities, Self-Weight, Imposed Loads for Buildings. The European Union: Brussels, Belgium, 2004.
39. *EN 1991-1-4*; Eurocode 1: Actions on Structures—Wind. Actions. The European Union: Brussels, Belgium, 2008.
40. *EN 1991-1-3*; Eurocode 1: Actions on Structures—Snow Loads. The European Union: Brussels, Belgium, 2005.
41. *EN 1995-1-1:2004/A2:2014*; Eurocode 5: Design of Timber Structures—Part. 1-1: General.—Common. Rules and Rules for Buildings. The European Union: Brussels, Belgium, 2014.
42. Renev, I.A.; Chechurin, L.S. Application of TRIZ in Building Industry: Study of Current Situation. In *Proceedings of the Procedia CIRP*; Elsevier B.V.: Amsterdam, The Netherlands, 2016; Volume 39, pp. 209–215.
43. Spreafico, C. Can TRIZ (Theory of Inventive Problem Solving) Strategies Improve Material Substitution in Eco-Design? *Sustain. Prod. Consum.* **2022**, *30*, 889–915. [CrossRef]
44. Palma, P.; Steiger, R. Structural Health Monitoring of Timber Structures—Review of Available Methods and Case Studies. *Constr. Build. Mater.* **2020**, *248*, 118528. [CrossRef]
45. López, G.; Vallelado-Cordobés, P.; Gómez-Royuela, J.L.; Basterra, L.A. Diagnosis and Assessment of a Historic Timber Structure in La Casa Del Corregidor, Using Non-Destructive Techniques. *Case Stud. Constr. Mater.* **2023**, *19*, e02311. [CrossRef]
46. Branco, J.M.; Sousa, H.S.; Tsakanika, E. Non-Destructive Assessment, Full-Scale Load-Carrying Tests and Local Interventions on Two Historic Timber Collar Roof Trusses. *Eng. Struct.* **2017**, *140*, 209–224. [CrossRef]
47. Machado, J.S.; Pereira, F.; Quilhó, T. Assessment of Old Timber Members: Importance of Wood Species Identification and Direct Tensile Test Information. *Constr. Build. Mater.* **2019**, *207*, 651–660. [CrossRef]
48. Skotnicka-Siepsiak, A. The Applicability of Coanda Effect Hysteresis for Designing Unsteady Ventilation Systems. *Energies* **2021**, *14*, 34. [CrossRef]
49. Skotnicka-Siepsiak, A. An Evaluation of the Performance of a Ground-to-Air Heat Exchanger in Different Ventilation Scenarios in a Single-Family Home in a Climate Characterized by Cold Winters and Hot Summers. *Energies* **2022**, *15*, 105. [CrossRef]
50. Conference Materials Technical Expert. In Proceedings of the Workshop (Warsztat Pracy Rzeczoznawcy Budowlanego), Cedzyna, Poland, 26–28 October 2020. (In Polish)
51. *EN 772-1:2011+A1:2015*; Methods of Test for Masonry Units—Part 1: Determination of Compressive Strength. The European Union: Brussels, Belgium, 2020.

materials

MDPI

Article

Description of Material Properties of Degraded and Damaged Segments of Multi-Leaf Masonry in Analyses of Large Three-Dimensional Structures

Czesław Miedziałowski and Adam Walendziuk *

Department of Building Structures and Structural Mechanics, Faculty of Civil Engineering and Environmental Sciences, Bialystok University of Technology, Wiejska 45E, 15-351 Białystok, Poland; c.miedzialowski@pb.edu.pl
* Correspondence: a.walendziuk@pb.edu.pl

Abstract: This article focuses on the description of material properties of segments of masonry structures in three-dimensional analyses. It mainly considers degraded and damaged multi-leaf masonry walls. In the beginning, the causes of degradation and damage to masonry are described with examples. It was reported that the analysis of such structures is difficult due to the adequate description of the mechanical properties in the individual segments of the structure and the amount of computational cost of large three-dimensional structures. Next, a method of describing large fragments of masonry structures by means of macro-elements was proposed. The formulation of such macro-elements in three-dimensional and two-dimensional problems was given by introducing limits of variation in material parameters and damage of structures expressed by the limits of integration of macro-elements with specified internal structures. Then, it was stated that such macro-elements can be used to build computational models by the finite element method, which allows the analysis of the deformation–stress state, and at the same time, reduce the number of unknowns in such issues. A strategy for performing analyses and examples of practical applications in masonry analyses were proposed. It was reported that the results of the analyses can be used to plan the repairs and strengthening of structures. Finally, the conducted considerations and proposals were summarised, as well as examples of practical applications.

Keywords: masonry; non-homogenous materials; degradation; damage; macro-element

Citation: Miedziałowski, C.; Walendziuk, A. Description of Material Properties of Degraded and Damaged Segments of Multi-Leaf Masonry in Analyses of Large Three-Dimensional Structures. *Materials* **2023**, *16*, 4076. https://doi.org/10.3390/ma16114076

Academic Editor: Krzysztof Schabowicz

Received: 14 March 2023
Revised: 25 May 2023
Accepted: 27 May 2023
Published: 30 May 2023

1. Introduction

A significant part of the materials used in the past as well as nowadays in the construction industry are heterogeneous materials and structures such as concrete, reinforced concrete, and masonry. Masonry as a construction material has been used for centuries. Unreinforced masonry can be defined as a composite of two interconnected components—masonry units and mortar. The classification (taxonomy) of the internal arrangements of masonry is extensive with regard to masonry made up of ceramic as well as stone elements and various layers (leaves) [1,2]. In addition to the strength of the components, the influence of an internal structure of the masonry (periodic, quasi-periodic, chaotic textures) on the formation of damage is still being explored. Especially for historical masonry structures, attention is also paid to structural arrangement and the internal structure of masonry walls [3]. Essential interrelated parameters of construction are the durability of component materials and material properties. These parameters are influenced by such factors as wall texture and its construction (single- or multi-leaf), unit shape and size, the volumetric ratio between components, and the mechanical properties of mortar and units—Young's modulus, Poisson's ratio, and compressive, tensile, and shear strength. Observations and research studies indicate the variability of these parameters over time [4–7]. A variety of environmental impacts affect embedded materials and buildings from their establishment.

During the realisation and utilisation of structures, physical, chemical, and biological interactions cause processes associated with unfavourable changes in the components, leading to gradual deterioration and degradation of properties of the materials used in structures. Damage due to overloading or frost effects can also occur during a lifetime. It is a natural process difficult to be stopped. A case in point is multi-leaf walls (including degraded ones), in the description of which should be distinguished segments in different states of technical condition. Examples of such degraded masonry structures are shown in Figure 1. Changes in stress states in structures are caused by the degradation of materials, deformations, changes in the purpose (function), and loads of the facility [8]. Degradation and ageing of materials and damages are influenced by the environment, including temperature effects, precipitation, or capillary rising of water from the subsoil. A change in the state of stress in the structure leads to damage in the form of discontinuities and cracks. In some geographical regions, water is an important factor affecting the durability of masonry. Freezing in the pores of the material leads to frost damage (flaking, spalling, or cracking) [9]. It also induces other corrosive processes, changes the texture of the brick, and allows the growth of microorganisms, leading to biodeterioration. Adverse impacts and changes in the properties of materials cause the need to assess the current strength condition and then protect them by increasing their load-bearing capacity and stiffness. Before any action is undertaken, it is important to recognise the scheme of working and stress distribution in a structure composed of masonry units and binders, considering the component arrangement and the mechanisms of damage.

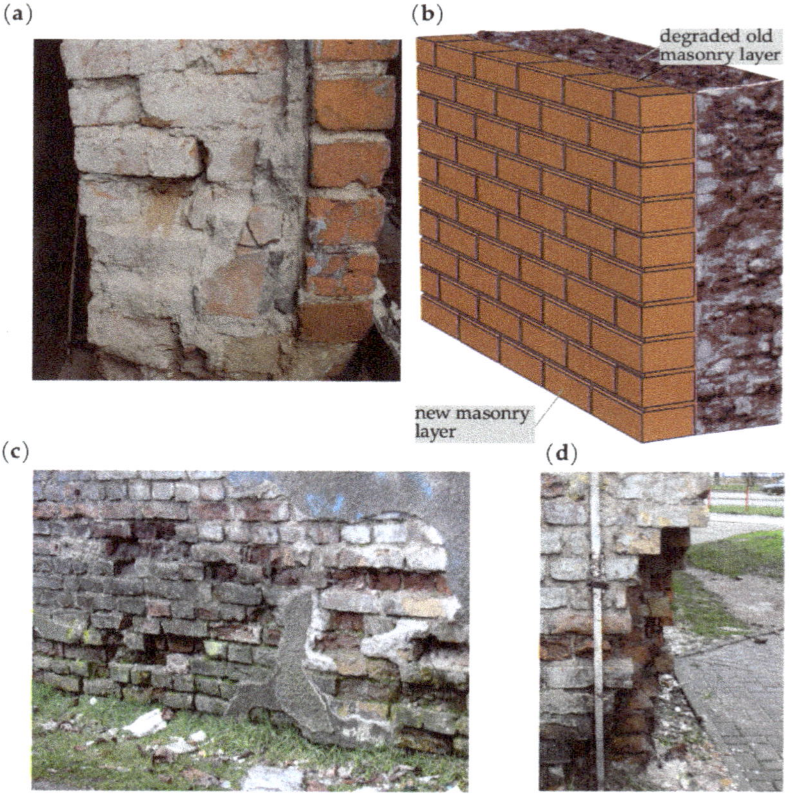

Figure 1. Examples of structures and degradation of masonry fragments (**a**–**d**).

The issues of materials joining are considered in engineering in many fields and aspects [10]. The analysis of heterogeneous materials investigates the interactions of the components and contact zones are also taken into account. These are studied at different scales of observation by assuming homogeneity (using homogenisation) or describing the internal arrangement of components in material. In numerical analyses of such materials, interface-type elements are used to describe the connections [11,12]. These have been used among others in modelling various material structures, for example, masonry [13] and concrete [14]. Material inhomogeneities are also introduced into structural materials by implementing modern repair methods through the injection or bonding of composite or steel materials to reinforced concrete or masonry elements. The scope also includes issues related to the renovation and strengthening of masonry structures and methods based on numerical analysis are useful during their implementation.

For the analysis of masonry structures, many numerical modelling methods have been developed to simulate the behaviour of masonry. In the most general approach, a distinction can be made between continuum models and discrete models. Continuum models are based mainly on the finite element method (FEM) [15]. Discrete models consider the medium as an assembly of distinct bodies interacting along the boundaries (DEM) [16]. A synthetic overview and comparison of the results of these methods can be found in the paper [17]. Depending on the level of accuracy of the description of the medium, the following modelling strategies are distinguished [18]: detailed micro-modelling, simplified micro-modelling, and macro-modelling. In detailed micro-modelling units and mortar, in joints are represented as continuous materials bonded by interface discontinuum elements. Simplified micro-modelling uses the geometric expansion of the units separated by discontinuous elements that simulate the behaviour of the mortar joints and unit-mortar interface. In macro-modelling masonry components: units, mortar, and unit–mortar interface are smeared out in a homogeneous continuum. The aforementioned approaches are based on the representation of the solid medium at the micro-, meso-, and macro-scale. Proposals for more detailed modelling strategy classifications in application to masonry structures are given in papers [19,20]. Among the methods currently being developed, particular attention is paid to multi-scale modelling by analysing the issue at several scales of observation [21], e.g., at the meso- and macro-scale.

Static analysis of masonry structures in spatial (3D) schemes needs the development of large-scale computational models [22]. The analysis of large real spatial structures in multi-year periods of utilisation and calculations in spatial schemes often require the considerable computing power of computers and the use of special computing procedures and techniques [23,24]. One way to solve this problem is through the multi-scale modelling of structures [25–27]. In order to reduce the computational cost, this paper proposes the use of multilayer macro-elements that model geometric and mechanical properties of fragments of structures. The proposed formulation of the description of segments (fragments) of the structure with differently localised zones with degradation and damage, as well as the multi-stage analysis, is filling the scientific and research gap. The description of masonry fragments can be regarded as a certain way of numerical homogenisation of masonry in a multi-scale approach [28]. The formulation of such elements and their properties is the main focus of this work. A proposal for a stepwise analysis and examples of the application of such a description to masonry structures will also be presented. Especially in the case of historical masonry walls, we encounter various forms of degradation and ways of implementing repair and strengthening [29–32]. A reliable description of the mechanism of static work of such masonry structures requires the use of spatial (3D) models, formulated, e.g., in the methodology of the finite element method.

2. Problem Formulation

In masonry structures, within the range of its technical condition considered, there are fragments in different levels of degradation and damage (Figure 2).

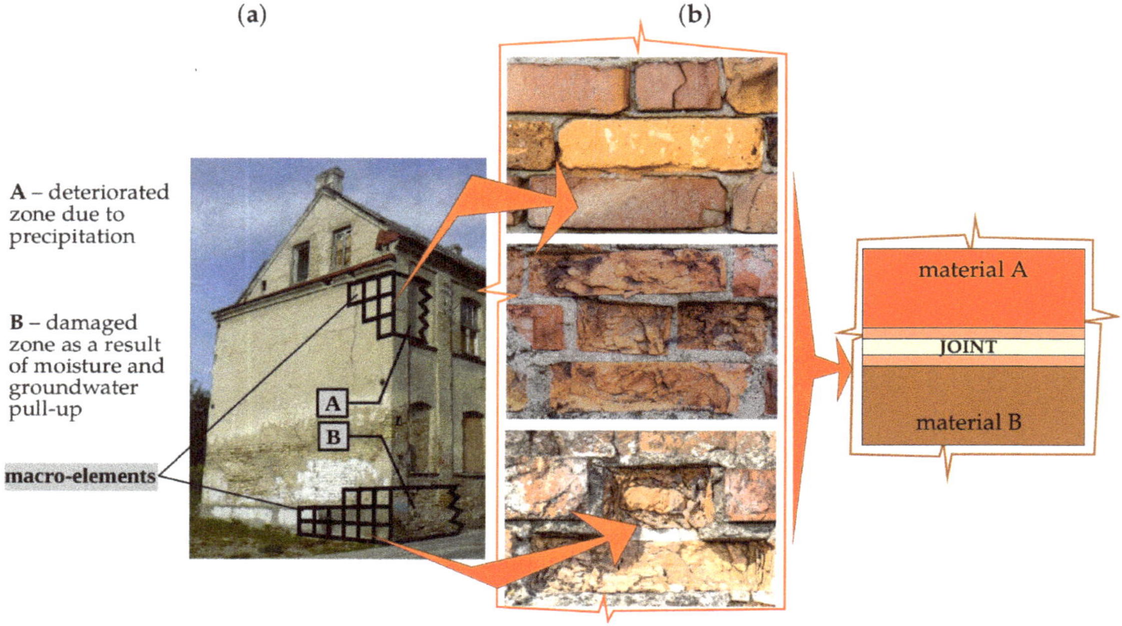

Figure 2. An example of a structure composed of heterogeneous and degraded materials: (**a**) different degradation zones in a building; (**b**) brick arrangement, local degradation forms, and meso-scale model.

Degradation is understood as a local change in the stiffness of the material, expressed through mass losses, cracking, surface flaking, and, as a consequence, changes in the mechanical parameters E, ν, R. Damage is treated as breaking a continuity of masonry and an exclusion of damaged areas or volumes from static work. The description of degradation and damage is analysed at two scales: at the structural scale, where an equivalent homogeneous medium is considered, and at the meso-scale, where the complex heterogeneous masonry mesostructure is taken into account. The developed method of description consists of separating segments of a certain size in the analysed structural model, referred to as macro-elements. In each macro-element, the existing (actual) state of degradation of the structure and damage in the form of discontinuity is assumed. As a result of integration in the sub-areas, matrices describing the stiffnesses of the macro-elements are determined. The stiffness of the whole structure is described by the set of all macro-elements.

A model presented in the work enables us to take into account the variation in physical properties appearing in sub-areas (sub-spaces) of a structure, the damaged fragments of a structure by introducing the limits of variation in material parameters, and the damage of a structure expressed by macro-element integration limits. The model describing larger areas (volumes) is proposed, allowing optimal modelling of the material properties of a structure in a three-dimensional scheme, e.g., by the finite element method, and avoiding high computational costs by using macro-elements and staged analyses. Depending on the dimensions of the analysed structure and the adopted discretisation method—the dimensions of repetitive masonry fragments—it is proposed to use 8- and 20-noded macro-elements, as shown in Figure 3. In this paper, considerations are limited to walls with a regular arrangement of masonry units.

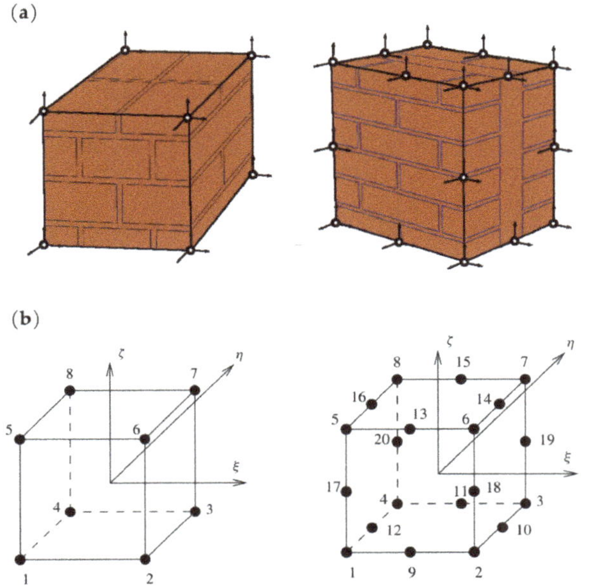

Figure 3. Proposed finite macro-elements describing segments of a structure: (**a**) internal structures of macro-elements in general cases; (**b**) 8- and 20-node finite elements.

In order to describe the phenomena occurring, areas covering materials with different properties are identified, assuming the continuity of displacements at the boundaries of these areas. The unknowns are located at the nodes and form a global displacement vector $\mathbf{q}$. In this vector, a number of displacements in the zone of the cohesion are selected and an element e is spanned on the nodes. The displacement components of the nodes of this element form a vector $\mathbf{u}_e$. In the analysis, it is proposed to use macro-elements that sample internal structures, as shown in Figure 3a. The finite macro-elements are spanned on 8 or 20 nodes, and the displacement components of the nodes of these elements have 24 and 60 components, respectively:

$$\mathbf{u}_e^{(8)} = \{u_1,\ v_1,\ w_1,\ \ldots,\ u_8,\ v_8,\ w_8\}^T, \tag{1}$$

$$\mathbf{u}_e^{(20)} = \{u_1,\ v_1,\ w_1,\ \ldots,\ u_{20},\ v_{20},\ w_{20}\}^T. \tag{2}$$

The displacement field in heterogeneous media, in which three directions of variation in physical properties of materials are observed, is assumed to be approximated by shape functions $N_i^{(e)}$. In the 8-node element, the polynomials approximating the displacement field are linear:

$$N_i^{(8)} = \tfrac{1}{8}(1+\xi\xi_i)(1+\eta\eta_i)(1+\xi\xi_i), \quad i = 1,\ldots,8 \tag{3}$$

whereas in the 20-node element these are second-degree polynomials [33] of the type:

$$N_i^{(20)} = \tfrac{1}{8}(1+\xi\xi_i)(1+\eta\eta_i)(1+\zeta\zeta_i)(\xi\xi_i+\eta\eta_i+\zeta\zeta_i-2), \quad i = 1,\ldots,8 \tag{4}$$

$$N_i^{(20)} = \tfrac{1}{4}(1-a^2)(1+bb_i)(1+cc_i), \quad a,b,c \in \{\eta,\xi,\zeta\}, \quad i = 9,\ldots,20 \tag{5}$$

where ξ_i, η_i, ζ_i are the natural coordinates of the node i.

The strain energy of the entire structure can be written by the equation:

$$\mathcal{W}_\epsilon = \frac{1}{2}\int_V \boldsymbol{\epsilon}^T \sigma dV = \frac{1}{2}\int_V \boldsymbol{\epsilon}^T \mathbf{E}\boldsymbol{\epsilon} dV = \frac{1}{2}\int_V (\mathbf{Bu})^T \mathbf{E}(\mathbf{Bu}) dV = \\ = \frac{1}{2}\mathbf{q}^T \sum_{e=1}^{elem} \left[\mathbf{R}_e^T \mathbf{K}_e \mathbf{R}_e\right] \mathbf{q} = \frac{1}{2}\mathbf{q}^T \mathbf{K}\mathbf{q}, \tag{6}$$

where:

- $\mathbf{E}$ —elasticity matrix;
- $\mathbf{B}$ —strain matrix;
- $\mathbf{R}_e$ —an allocation (incidence) matrix, assigns degrees of freedom to elements;
- $\mathbf{K}_e$ —macro-element stiffness matrix.

The macro-element stiffness matrix is calculated by integrating the expression for the internal strain energy in sub-spaces (sub-areas). Sample macro-element structures in spatial and flat problems are shown in Figure 4.

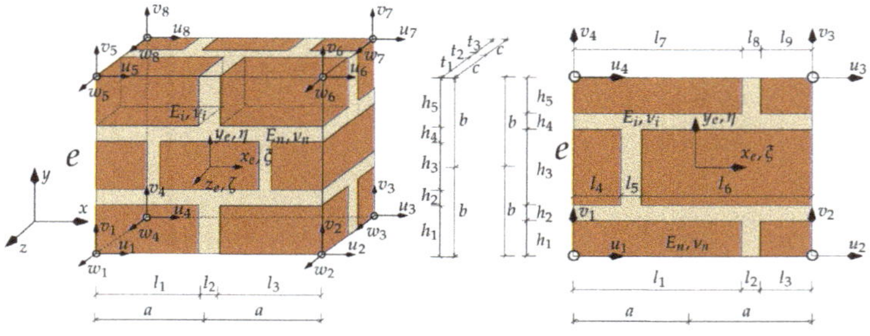

Figure 4. Sample internal structures of macro-elements with layers in three and two directions.

In spatial issues:

$$\mathbf{K}_e = \int_{V_1} \mathbf{B}^T \mathbf{E}^{(1)} \mathbf{B}\, dV + \ldots + \int_{V_j} \mathbf{B}^T \mathbf{E}^{(j)} \mathbf{B}\, dV + \ldots + \int_{V_n} \mathbf{B}^T \mathbf{E}^{(n)} \mathbf{B}\, dV. \tag{7}$$

Sub-areas of integration are fragments of materials with specific characteristics used in the construction of the wall. Figure 5 shows examples of sub-spaces in a wall with a thickness of two bricks and a wall with a three-leaf structure.

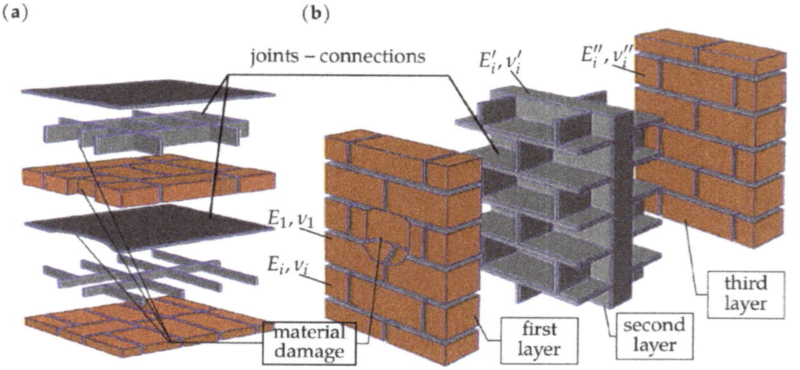

Figure 5. Integration sub-spaces of macro-element with damage: (**a**) repetitive layers (courses) of solid masonry; (**b**) multi-leaf masonry.

Numerical integration is carried out according to the relationship written in the form:

$$\mathbf{K}_e = \int_{-a}^{-a+l_1} dx \int_{-b}^{-b+h_1} dy \int_{-c}^{-c+t_1} \mathbf{B}^T \mathbf{E}^{(1)} \mathbf{B}\, dz + \\ + \int_{-a+l_1}^{-a+l_1+l_2} dx \int_{-b}^{-b+h_1} dy \int_{-c}^{-c+t_1} \mathbf{B}^T \mathbf{E}^{(2)} \mathbf{B}\, dz + \\ \cdots + \int_{-a+\sum l_{i-1}}^{-a+\sum l_i} dx \int_{-b+\sum h_{i-1}}^{-b+\sum h_i} dy \int_{-c+\sum t_{i-1}}^{-c+\sum t_i} \mathbf{B}^T \mathbf{E}^{(i)} \mathbf{B}\, dz + \\ \cdots + \int_{a-\sum l_{n-1}}^{a} dx \int_{b-\sum h_{i-1}}^{b} dy \int_{c-\sum t_{n-1}}^{c} \mathbf{B}^T \mathbf{E}^{(n)} \mathbf{B}\, dz \tag{8}$$

in which the symbols l, h, t are shown in Figure 4. The general form of the deformation matrix **B** (the derivatives of the shape function) is constant in each sub-area:

$$\mathbf{B} = \left[\mathbf{B}_1^{(e)},\ \ldots,\ \mathbf{B}_i^{(e)},\ \ldots,\ \mathbf{B}_n^{(e)}\right], \tag{9}$$

$$\mathbf{B}_i^{(e)} = \mathbf{L}\mathbf{N}_i^{(e)} = \mathbf{L}\begin{bmatrix} N_i^{(e)} & 0 & 0 \\ 0 & N_i^{(e)} & 0 \\ 0 & 0 & N_i^{(e)} \end{bmatrix}, \tag{10}$$

i —node number $(i = 1 \ldots n)$,
n —number of nodes of the macro-element,
L —operator matrix,
$N_i^{(e)}$—shape functions.

The elasticity matrix in sub-space j (bricks, joints), in general varying over time (T), is written in the form:

$$\mathbf{E}_j^{(T)} = \mathbf{E}_j = \frac{E_j}{(1+\nu_j)(1-2\nu_j)}\begin{bmatrix} (1-\nu_j) & \nu_j & \nu_j & 0 & 0 & 0 \\ \nu_j & (1-\nu_j) & \nu_j & 0 & 0 & 0 \\ \nu_j & \nu_j & (1-\nu_j) & 0 & 0 & 0 \\ 0 & 0 & 0 & \frac{(1-2\nu_j)}{2} & 0 & 0 \\ 0 & 0 & 0 & 0 & \frac{(1-2\nu_j)}{2} & 0 \\ 0 & 0 & 0 & 0 & 0 & \frac{(1-2\nu_j)}{2} \end{bmatrix}. \tag{11}$$

In the notation of matrix **E**, the superscript (T) is used to emphasise the variability of the parameters over time, but for the sake of clarity, it will be omitted in most notations, as in the notations of other matrices and vectors: **K**, **q**, **Q**. Elements of the matrix **E** dependent on the material constants, Young's modulus E, and Poisson's ratio ν of the brick and mortar are generally variable in the adopted sub-spaces j and during long-term service life of the structure, and should be determined in laboratory tests [34] (Figure 6).

(**a**) (**b**)

Figure 6. Laboratory testing of materials: (**a**) bricks; (**b**) mortar.

Flat models can be used in some cases. In a 2D area composed of sub-areas (Figure 7), integration in sub-area j, under certain assumptions, can be carried out analytically, obtaining the explicit form of the macro-element stiffness matrix. The following results are then obtained:

$$\begin{aligned}\mathbf{K}_e^{(T)} = \mathbf{K}_e = t\int_{-a}^{-a+l_1} dx \int_{-b}^{-b+h_1} \mathbf{B}^T\mathbf{E}^{(1)}\mathbf{B}\, dy + \\ + t\int_{-a+l_1}^{a-l_3} dx \int_{-b}^{-b+h_1} \mathbf{B}^T\mathbf{E}^{(2)}\mathbf{B}\, dy + \cdots \\ \cdots + t\int_{-a+l_9}^{a-l_{11}} dx \int_{b-h_5}^{b} \mathbf{B}^T\mathbf{E}^{(10)}\mathbf{B}\, dy + t\int_{a-l_{11}}^{a} dx \int_{b-h_5}^{b} \mathbf{B}^T\mathbf{E}^{(11)}\mathbf{B}\, dy,\end{aligned} \tag{12}$$

and t denotes the layer thickness.

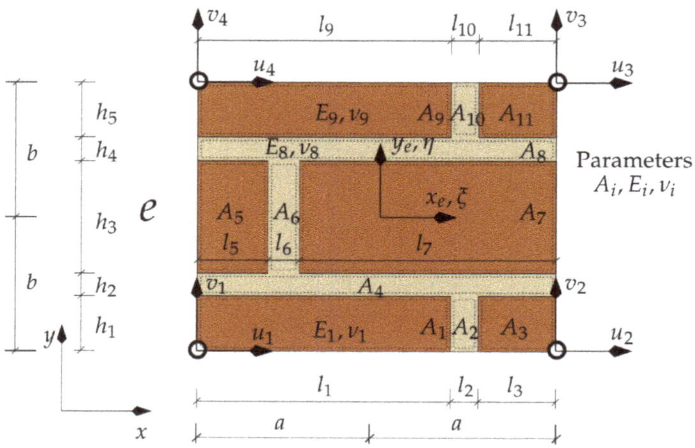

Figure 7. A sample internal structure of thickness-constant macro-element.

Below is enclosed a piece of the code for calculating the first element of the stiffness matrix obtained by integration of expression (12). The selected terms of the sum are equal:

```
KA(1,1,1) = Em(1)*h(1)*l(1)*(l(1)**2*ni(1)-6*a*l(1)*ni(1)+12*a**2*ni(1) &
            -l(1)**2+6*a*l(1)-2*h(1)**2+12*b*h(1)-24*b**2-12*a**2) &
            /(a**2*b**2*(ni(1)-1)*(ni(1)+1)),

  ⋮

KA(11,1,1) = Em(11)*h(5)*l(11)*(l(11)**2*ni(11)-l(11)**2-2*h(5)**2) &
             /(a**2*b**2*(ni(11)-1)*(ni(11)+1)).
```

The value of the element (1,1) of the stiffness matrix is then calculated from the formula:

```
Ke2D(1,1) = t/96 * ( KA(1,1,1) + KA(2,1,1)  + KA(3,1,1) + KA(4,1,1) &
                   + KA(5,1,1) + KA(6,1,1)  + KA(7,1,1) + KA(8,1,1) &
                   + KA(9,1,1) + KA(10,1,1) + KA(11,1,1) ).
```

3. Strategy for Performance of Analyses

It is proposed that the calculations should be carried out according to a staged algorithm that uses the formulation of the equilibrium equations in the form:

$$\mathbf{K}_i^{(T)}\mathbf{q}_i^{(T)} = \mathbf{Q}_i^{(T)}, \tag{13}$$

where:

$\mathbf{K}_i^{(T)}$ is variable over time stiffness matrix of the modelled medium;
$\mathbf{q}_i^{(T)}$ is the vector of unknown node displacements of the model at step i of the iteration;
$\mathbf{Q}_i^{(T)}$ is the load vector.

The formulation makes it possible to model material degradation and damage in segments of large existing structures or to investigate the effects of loading history over years of service on the degradation of masonry and the state of stress level in a structure. The computer implementation of the algorithm takes advantage of the secant stiffness matrix, seeking equilibrium of the system at each load increment. The scheme for carrying out the calculations has been divided into stages, as illustrated in Figure 8.

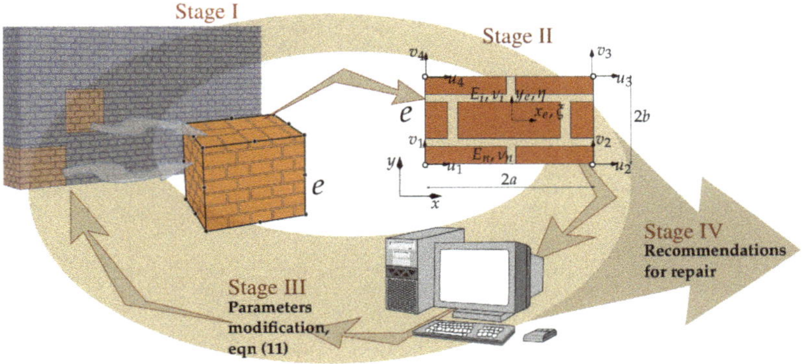

Figure 8. Stages of numerical analysis of heterogeneous materials embedded into a construction.

Stage I —analysis of large areas or volumes and macro-element formation;
Stage II —analysis of displacements, assessment of stress and strain fields inside the macro-elements;
Stage III—parameters modification in a model according to (11) and recomputation;
Stage IV—recommendations for repairs and strengthening.

As stated above, the method of analysis developed consists of separating segments (fragments) of a certain size in the structural model to be analysed, taking into account the unit arrangement and degradation. The unit arrangement (texture) does not need to be regular or repeatable; however, this affects the cost of the computations. For each segment, a specified (existing) state of degradation of the internal structure and the properties of the materials that compose it is assumed, in addition to possible damages and discontinuities of the structure. As a result of the integration in the sub-spaces (sub-areas) of the segments, the matrices used to model the stiffness of the segments (macro-element stiffness matrices) are calculated. The set of all such macro-elements forms the stiffness matrix of the modelled structure $\mathbf{K}_i$ in successive steps of load increments $\mathbf{Q}_i$. The load increment can also be interpreted as a time-dependent parameter. Once the displacement field of the model nodes has been determined, the stress state in each sub-area is verified. When the stresses exceed the permissible values, the current damage configuration of the internal structure is determined in each macro-element by searching for an equilibrium configuration at the meso-scale (at the RVE level). For the modified RVE structure, the macro-element stiffness matrix is calculated for the next loading step. The implementation of this calculation procedure is presented in a simplified flowchart (Figure 9), where only the most essential steps of the computational process are indicated. The flowchart is directly related to the numerical analysis strategy presented in Figure 8. It also contains graphical information consistent with the approach presented in Section 2 of this article describing degradation and damage processes. The masonry wall structures analysed are treated as brittle and a simple approach was used in the computational process that does not directly take into account other complex aspects, including that of plasticity. Exceeding the permissible stresses in brittle structures results in the elimination of the area of the structure where damage propagation is identified. This is expressed in a change in the integration limits or modification of the properties of the fragment of the internal structure of the macro-element, as illustrated in Figure 9. The proposed approach employs therefore an isotropic damage model of the internal structure of the macro-element.

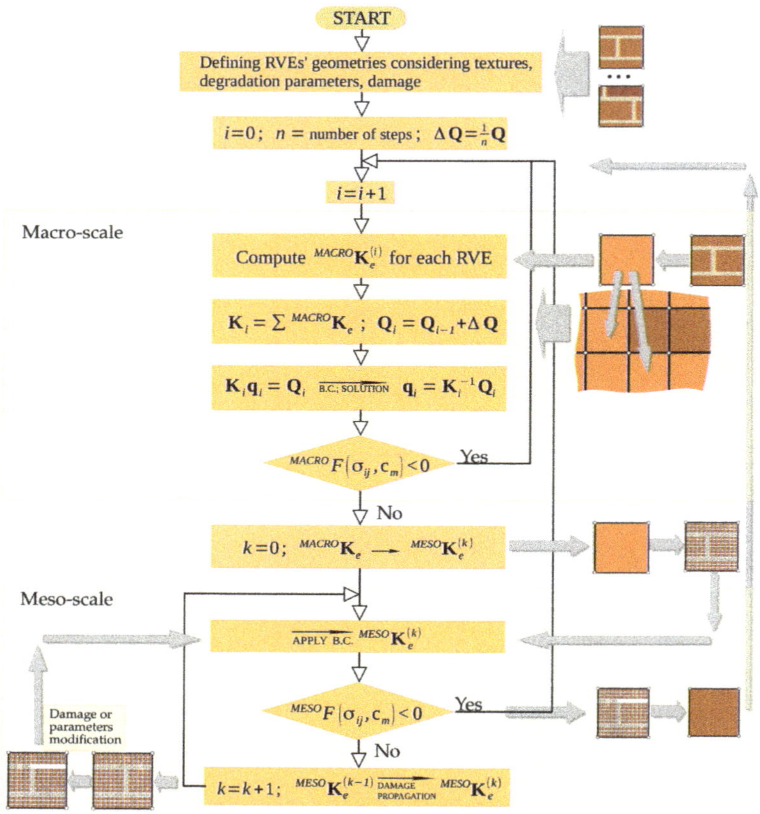

Figure 9. Flowchart of step-by-step computational process.

In order to perform stress assessment, determine damage zones in heterogeneous materials and structures, as well as to identify damage propagation and degradation mechanisms during the loading process, failure criteria are used [35]. They are formulated analytically in the form of functions of stresses, strains, or their invariants using the results of laboratory strength tests. The safe state of the stress field σ_{ij} is defined by the inequality:

$$F(\sigma_{ij}, c_m) < 0, \tag{14}$$

and c_m denotes the material parameters obtained in laboratory tests. The failure initiation of non-symmetric materials in the range of tensile stresses is verified by the Rankine criterion. The failure of brittle materials also depends on the value of the mean stress. Among the numerous failure criteria used in masonry analyses, the Mohr–Coulomb criterion is often applied [36,37]. In the present work, the criterion is adopted in the form:

$$\frac{1}{2\cos\varphi}(\sigma_i - \sigma_j) + \frac{1}{2}(\sigma_i + \sigma_j)\tan\varphi < R_s, \quad i,j = 1,2,3 \tag{15}$$

where φ is the angle of internal friction.

The principal stresses σ_i, σ_j are determined in macro-element structures from the computed displacements of its nodes. Additionally, the criterion is used to control the state of stress level in joints in which the physical plane of stress is determined, and it is written by an inequality:

$$|\tau| = c - \sigma_n \tan\varphi < R_s, \tag{16}$$

in which R_s is the shear strength of the joint.

4. Application Examples

The applicability of the model is demonstrated by examples. They were performed according to the presented step-by-step algorithm for successive load increments in order to reproduce the results of laboratory tests in which degradation over a longer period of time is unimportant. The first example presents an analysis of a fragment of masonry shearing parallel to the load-bearing joints with the geometry given in Figure 10. The discretisation with 324 macro-elements resulted in the division is shown in Figure 10a. The comparative discretisation by standard plane stress elements contains 16,125 finite elements in the size of 1 cm × 1 cm. The macro-elements used have dimensions several times larger, with side lengths in the range of 6–10 cm presented with lines in black. Material parameters were assumed to be the same in the brick and mortar areas, respectively, (degradation and damage of the initial masonry structure were not taken into account). The number of unknowns in the developed models was 722 using macro-elements and 5942 using conventional modelling, respectively. Subsequent calculation steps were carried out according to the algorithm developed and shown in the flowchart. As a result of the solution, the damage distribution in the masonry sample was obtained. An exceedance of the stress values according to the adopted failure criterion was interpreted as material damage, as illustrated in Figure 10b, and compared with a typical form of damage (Figure 10c). Typical forms of damage obtained in laboratory tests of this type of masonry specimens can be found in papers [38,39], among others.

(a) (b) (c)

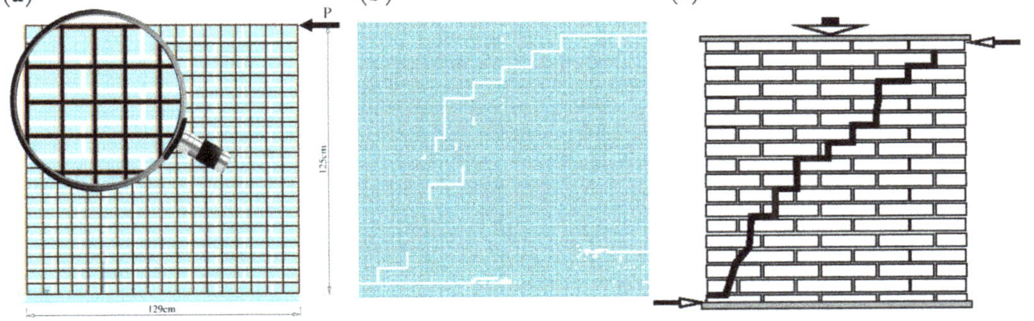

Figure 10. Shear wall: (**a**) model discretisation; (**b**) damage propagation pattern according to the adopted algorithm; (**c**) a form of typical failure mechanism in masonry under shear and compression.

The qualitative similarity of the obtained results of the two presented tests and a consistent trend in the form of damage to the masonry sample are observed. In both presented examples of numerical analyses, the same values of parameters characterising ceramic bricks and mortar were used: Young's modulus, Poisson's ratio, compressive strength, tensile strength, and shear strength were equal, respectively, to:

$$E^{(brick)} = 17,500 \text{ MPa}, \qquad E^{(mortar)} = 2900 \text{ MPa},$$
$$\nu^{(brick)} = 0.20, \qquad \nu^{(mortar)} = 0.20,$$

$$R_c^{(brick)} = 27.1 \text{ MPa}, \qquad R_c^{(mortar)} = 10.0 \text{ MPa},$$
$$R_t^{(brick)} = \tfrac{1}{7} R_c^{(brick)}, \qquad R_t^{(mortar)} = \tfrac{1}{5} R_c^{(mortar)},$$
$$R_s^{(brick)} = \tfrac{4}{5} R_t^{(brick)}, \qquad R_s^{(mortar)} = \tfrac{4}{5} R_t^{(mortar)}.$$

A further example refers to a part of brick masonry representing a flexural member of a structure strengthened with an FRP strip placed under the lower course of bricks. The arrangement of the units in the brickwork element studied and the macro-element meshing discretising the masonry units and mortar are shown in Figure 11. A total of 490 macro-elements with side dimensions of 3–5 cm and 7611 elements of the classical FEM discretisation were used. The number of unknowns in the model where macro-elements were used was 1080, compared to 7280 unknown displacements in the classical model. In the numerical experiment, a linear model of the strengthening material was adopted with a value of Young's modulus E^{FRP} = 240 GPa. The FRP strip material was modelled with bar finite elements, assuming no damage between the strengthening layer and the bricks. The numerical experiment was carried out according to the adopted algorithm in the flowchart. The presented damage form, similarly as in the first example, was determined on the basis of the adopted failure criterion.

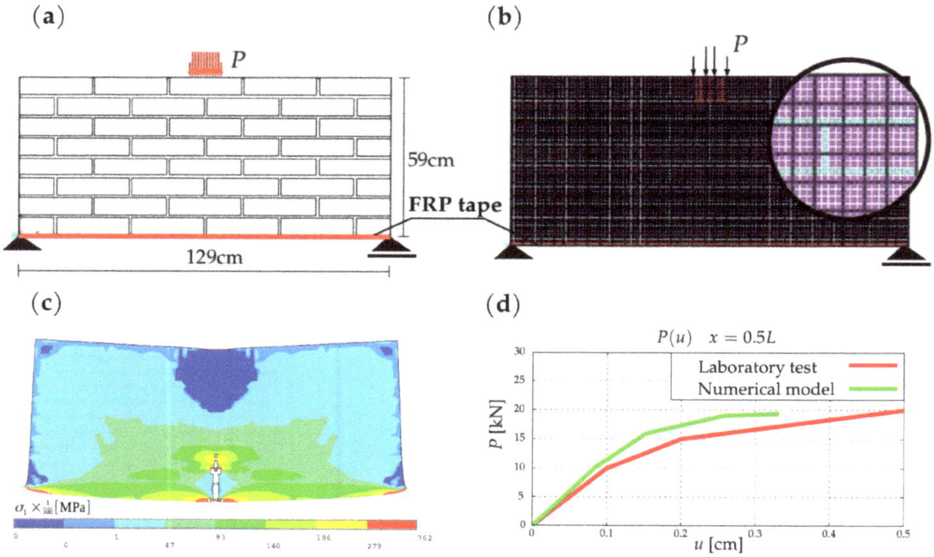

Figure 11. The model analysed: (**a**) scheme and geometry; (**b**) discretisation of the masonry structure; (**c**) the stresses in the masonry; (**d**) the $P(u)$ force–displacement relationship diagram.

The realisation of the real element test is described in the work [40]. The results of the strength laboratory experiments and numerical simulation are shown in Figure 11c,d. The displacement–force relationship graphs obtained in both experiments show convergence. Analysing the differences in the obtained correlations for the two experiments, these were determined to be 16.6% for 0.1 cm displacements, 11.5% for 0.2 cm displacements, and 8.2% for 0.3 cm displacement values. Comparisons of the results show a convergent trend in both experiments.

5. Conclusions

The work presents proposals for the description and analysis of heterogeneous structures, composed of dissimilar materials in different states of technical condition and level of degradation. Examples include masonry structures of historical buildings and other masonry structures. The demonstrated method of modelling with macro-elements allows the analysis of large fragments of structures, taking into account their heterogeneity and degradation over time. Stiffness matrices of the macro-element describing such fragments were obtained by numerical or explicit (analytical) integration within the limits of variation in material parameters, degradation, and damage to the structure. This method can be regarded as a kind of numerical homogenisation of the structural arrangement of masonry structures. In this work, such elements were formulated and application examples were given, along with a proposal for staged analysis of structures. The results obtained in the examples show the correct modelling of structures using the proposed method and macro-elements, show the correct tendency of relationships, and quite good convergence of computational and experimental results varying within 8–16%. The modelling method and strategy make it possible to analyse structures in spatial (3D) or flat (2D) schemes and reduce the number of unknowns by about seven times. In engineering practice, there is a need for the repair and rehabilitation of heterogeneous and degraded structures. The modelling approach given in the work allows the assessment of the stress state in structural elements in order to perform the repairs and strengthening of multi-material structures, as illustrated in Figure 11.

Author Contributions: Conceptualisation, C.M. and A.W.; methodology, C.M. and A.W.; software, A.W.; validation, C.M. and A.W.; formal analysis, C.M. and A.W.; investigation, C.M. and A.W.; resources, C.M. and A.W.; writing—original draft preparation, A.W.; writing—review and editing, C.M. and A.W.; visualisation, A.W.; supervision, C.M. and A.W.; project administration, C.M.; funding acquisition, C.M. All authors have read and agreed to the published version of the manuscript.

Funding: The research was carried out within the scope of work no. WZ/WB-IIL/4/2023 and financed from the resources for science of Ministry of Education and Science of Poland.

Institutional Review Board Statement: Not applicable.

Informed Consent Statement: Not applicable.

Data Availability Statement: Data sharing is not applicable to this article.

Conflicts of Interest: The authors declare no conflict of interest.

References

1. Szabó, S.; Funari, M.F.; Lourenço, P.B. Masonry patterns' influence on the damage assessment of URM walls: Current and future trends. *Dev. Built Environ.* **2023**, *13*, 100119. . [CrossRef]
2. Schiantella, M.; Cluni, F.; Gusella, V. Parametric Analysis of Failure Loads of Masonry Textures by Means of Discontinuity Layout Optimization (DLO). *Materials* **2022**, *15*, 3691. . [CrossRef] [PubMed]
3. Rios, A.J.; Pingaro, M.; Reccia, E.; Trovalusci, P. Statistical assessment of in-plane masonry panels using limit analysis with sliding mechanism. *J. Eng. Mech.* **2022**, *148* . [CrossRef]
4. Ramalho, M.; Taliercio, A.; Anzani, A.; Binda, L.; Papa, E. Experimental and numerical study of multi-leaf masonry walls. In *Structural Studies, Repairs and Maintenance of Heritage Architecture IX*; Brebbia, C.A., Torpiano, A., Eds.; WIT Press: Southampton, UK, 2005; pp. 333–342.
5. Anzani, A.; Garavaglia, E.; Binda, L. Long-term damage of historic masonry: A probabilistic model. *Constr. Build. Mater.* **2009**, *23*, 713–724. [CrossRef]
6. Miedziałowski, C. Corrosion and stress damage of facade elements of historic buildings. In *Protection of Buildings from Moisture, Biological Corrosion and Fire. Vol. 14*; Skowroński, W., Ed.; Polish Association of Building Mycologists: Wrocław, Poland, 2017; pp. 125–136. (In Polish)
7. Borri, A.; Corradi, M.; Castori, G.; De Maria, A. A method for the analysis and classification of historic masonry. *Bull. Earthq. Eng.* **2015**, *13*, 2647–2665. [CrossRef]
8. Krentowski, J.; Chyży, T.; Dunaj, P. Sudden collapse of a 19th-century masonry structure during its renovation process. *Eng. Fail. Anal.* **2017**, *82*, 540–553. [CrossRef]

9. Stryszewska, T.; Kańka, S. Forms of Damage of Bricks Subjected to Cyclic Freezing and Thawing in Actual Conditions. *Materials* **2019**, *12*, 1165. [CrossRef]
10. Messler, R.W., Jr. *Joining of Materials and Structures. From Pragmatic Process to Enabling Technology*; Elsevier, Butterworth-Heinemann: Oxford, UK, 2004.
11. Desai, C.S.; Zaman, M.M.; Lightner, J.G.; Siriwardane, H.J. Thin-layer element for interfaces and joints. *Int. J. Numer. Anal. Methods Geomech.* **1984**, *8*, 19–43. [CrossRef]
12. Hu, L.; Pu, J.L. Application of damage model for soil–structure interface. *Comput. Geotech.* **2003**, *30*, 165–183. [CrossRef]
13. Lourenço, P.B.; Rots, J.G. A multi-surface interface model for the analysis of masonry structures. *J. Eng. Mech.* **1997**, *123*, 660–668. [CrossRef]
14. Willam, K.; Rhee, I.; Shing, B. Interface damage model for thermomechanical degradation of heterogeneous materials. *Comput. Methods Appl. Mech. Eng.* **2004**, *193*, 3327–3350. [CrossRef]
15. Zienkiewicz, O.C.; Taylor, R.L.; Zhu, J.Z. *The Finite Element Method: Its Basis and Fundamentals*, 7th ed.; Elsevier, Butterworth-Heinemann: Oxford, UK, 2013.
16. Lemos, J. Discrete element modeling of masonry structures. *Int. J. Archit. Herit. Conserv. Anal. Restor.* **2007**, *1*, 190–213. . [CrossRef]
17. Baraldi, D.; Reccia, E.; Cecchi, A. In plane loaded masonry walls: DEM and FEM/DEM models. A critical review. *Meccanica* **2018**, *53*, 1613–1628. [CrossRef]
18. Lourenço, P.B. Computations of historical masonry constructions. *Prog. Struct. Eng. Mater.* **2002**, *4*, 301–319. [CrossRef]
19. Asteris, P.G.; Plevris, V.; Sarhosis, V.; Papaloizou, L.; Mohebkhah, A.; Komodromos, P.; Lemos, J.V. Numerical modeling of historic masonry structures. In *Handbook of Research on Seismic Assessment and Rehabilitation of Historic Structures*; Asteris, P.G., Plevris, V., Eds.; IGI Global: Hershey, PA, USA, 2015; pp. 213–255.
20. D'Altri, A.M.; Sarhosis, V.; Milani, G.; Rots, J.; Cattari, S.; Lagomarsino, S.; Sacco, E.; Tralli, A.; Castellazzi, G.; de Miranda, S. Modeling strategies for the computational analysis of unreinforced masonry structures: Review and classification. *Arch. Comput. Methods Eng.* **2020**, *27*, 1153–1185. [CrossRef]
21. Sacco, E.; Addessi, D.; Sab, K. New trends in mechanics of masonry. *Meccanica* **2018**, *53*, 1565–1569. [CrossRef]
22. Roca, P.; Cervera, M.; Gariup, G.; Pela, L. Structural analysis of masonry historical constructions. Classical and advanced approaches. *Arch. Comput. Methods Eng.* **2010**, *17*, 299–325. [CrossRef]
23. Niekamp, R.; Markovic, D.; Ibrahimbegovic, A.; Matthies, H.G.; Taylor, R.L. Multi-scale modelling of heterogeneous structures with inelastic constitutive behavior. Part II – software coupling implementation aspects. *Eng. Comput.* **2009**, *26*, 6–28. [CrossRef]
24. Ibrahimbegovic, A.; Davene, L.; Colliat, J.-B.; Brancherie, D. Computational mechanics of integrity & durability in extreme environment for concrete & reinforced-concrete structures. *IACM Expr.* **2010**, *27*, 9–15.
25. Leonetti, L.; Greco, F.; Trovalusci, P.; Luciano, R.; Masiani, R. A multiscale damage analysis of periodic composites using a couple-stress/Cauchy multidomain model: Application to masonry structures. *Compos. B Eng.* **2018**, *141*, 50–59. [CrossRef]
26. Lourenço, P.B.; Silva, L.C. Computational applications in masonry structures: From the meso-scale to the super-large/super-complex. *Int. J. Multiscale Comput. Eng.* **2020**, *18*, 1–30. [CrossRef]
27. Drougkas, A.; Sarhosis, V. Micro-mechanical homogenisation of three-leaf masonry walls under compression. *Eng. Struct.* **2021**, *245* , 112890. [CrossRef]
28. Geers, M.G.D.; Kouznetsova, V.G.; Matouš, K.; Yvonnet, J. Homogenization methods and multiscale modeling: Nonlinear problems. In *Encyclopedia of Computational Mechanics, Part 1. Solids and Structures*, 2nd ed.; John Wiley & Sons: Hoboken, NJ, USA, 2017; pp. 1–34.
29. Kutut, V.; Ustinovičius, L. *Systemotechnical Assessment of the Real Cultural Heritage Management*; Publishing House of Bialystok University of Technology: Białystok, Poland, 2021; pp. 22–29.
30. Roca, P. Considerations on significance of history for the structural analysis of anscient constructions. In Proceedings of the IVth Int. Seminar on Structural Analysis of Historical Constructions, Padwa, Italy, 10–13 November 2004; pp. 63–73.
31. Modena, C. Design aproaches of interventions for the safety and conservation of historic buildings. In Proceedings of the IVth Int. Seminar on Structural Analysis of Historical Constructions, Padwa, Italy, 10–13 November 2004; pp. 75–83.
32. Jasieńko, J.; Łodygowski, T.; Rapp, P. *Repair, Maintenance and Strengthening of Selected Historic Brick Structures*; Dolnośląskie Wydawnictwo Edukacyjne: Wrocław, Poland, 2006. (In Polish)
33. Smith, I.M.; Griffiths, D.V.; Margetts, L. *Programming the Finite Element Method*, 5th ed.; John Wiley & Sons: Chichester, UK, 2014; pp. 615–617.
34. Jasieńko, J.; Szyszka, M. Assessing mechanical parameters of historical masonry structures through experimental testing in relation to structural conservation work. *J. Herit. Conserv.* **2013**, *36*, 7–17.
35. Ottosen, N.S.; Ristinmaa, M. *The Mechanics of Constitutive Modeling*; Elsevier: Oxford, UK, 2005.
36. Milani, G.; Lourenço, P.B.; Tralli, A. Homogenised limit analysis of masonry walls, Part I: Failure surfaces, *Comput. Struct.* **2006**, *84*, 166–180. [CrossRef]
37. Milani, G.; Lourenço, P.B.; Tralli, A. Homogenised limit analysis of masonry walls, Part II: Structural examples. *Comput. Struct.* **2006**, *84*, 181–195. [CrossRef]
38. Mann, W.; Müller, H. Failure of shear-stressed masonry–An enlarged theory, tests and application to shear-walls. *Proc. Br. Ceram. Soc.* **1982**, *30*, 223–235.

39. Dolatshahi, K.M.; Aref, A.J. Two-dimensional computational framework of meso–scale rigid and line interface elements for masonry structures. *Eng. Struct.* **2011**, *33*, 3657–3667. [CrossRef]
40. Walendziuk, A. Computer Simulation of Failure States Changes in Brittle Heterogeneous Materials and Structures Induced by External Interference Processes. Ph.D. Thesis, Faculty of Civil and Environmental Engineering, Bialystok University of Technology, Białystok, Poland, 2016. (In Polish)

materials

MDPI

Article

Fused Filament Fabrication and Computer Numerical Control Milling in Cultural Heritage Conservation

Daniela Fico [1,*], Daniela Rizzo [2], Francesco Montagna [1] and Carola Esposito Corcione [1]

[1] Department of Engineering for Innovation, University of Salento, Edificio P, Campus Ecotekne, s.p. 6 Lecce-Monteroni, 73100 Lecce, Italy
[2] Department of Cultural Heritage, University of Salento, Via D. Birago 64, 73100 Lecce, Italy
* Correspondence: daniela.fico@unisalento.it

Abstract: This paper reports a comparison between the advantages and disadvantages of fused filament fabrication (FFF) and computer numerical control (CNC) milling, when applied to a specific case of conservation of cultural heritage: the reproduction of four missing columns of a 17th-century tabernacle. To make the replica prototypes, European pine wood (the original material) was used for CNC milling, while polyethylene terephthalate glycol (PETG) was used for FFF printing. Neat materials were chemically and structurally characterized (FTIR, XRD, DSC, contact angle measurement, colorimetry, and bending tests) before and after artificial aging, in order to study their durability. The comparison showed that although both materials are subject to a decrease in crystallinity (an increase in amorphous bands in XRD diffractograms) and mechanical performance with aging, these characteristics are less evident in PETG ($E = 1.13 \pm 0.01$ GPa and $\sigma = 60.20 \pm 2.11$ MPa after aging), which retains water repellent ($ca = 95.96 \pm 5.56°$) and colorimetric ($\Delta E = 2.6$) properties. Furthermore, the increase in flexural strain (%) in pine wood, from $3.71 \pm 0.03\%$ to $4.11 \pm 0.02\%$, makes it not suitable for purpose. Both techniques were then used to produce the same column, showing that for this specific application CNC milling is quicker than FFF, but, at the same time, it is also much more expensive and produces a huge amount of waste material compared to FFF printing. Based on these results, it was assessed that FFF is more suitable for the replication of the specific column. For this reason, only the 3D-printed PETG column was used for the subsequent conservative restoration.

Keywords: advanced technologies; cultural heritage; conservation; additive manufacturing; computer numerical control milling

Citation: Fico, D.; Rizzo, D.; Montagna, F.; Esposito Corcione, C. Fused Filament Fabrication and Computer Numerical Control Milling in Cultural Heritage Conservation. *Materials* **2023**, *16*, 3038. https://doi.org/10.3390/ma16083038

Academic Editor: Krzysztof Schabowicz

Received: 19 March 2023
Revised: 7 April 2023
Accepted: 10 April 2023
Published: 12 April 2023

1. Introduction

Additive Manufacturing (AM), first named rapid prototyping before becoming known as "3D printing", is an innovative technology able to produce 3D models with complex shapes using a layer-by-layer building strategy. By contrast, in recognized traditional techniques, such as subtractive manufacturing (SM) and formative manufacturing (FM), the material is removed via machining, drilling, or grinding methods, or cast into molds [1]. Due to its fast expansion, the aim of AM has shifted from rapid prototyping to rapid tooling, and now to 3D manufacturing. AM has included several manufacturing techniques (photopolymerization, powder bed fusion, directed energy deposition, material extrusion, binder jetting, curing, lamination, etc.), to create a wide range of machineries of possible importance to industry. The driving force of the exponential increase in AM technologies could be attributed to the important results achieved by academic and industrial researchers in developing low-cost technologies and innovative high-tech materials [2–8]. Numerous aspects inform the selection of the most suitable manufacturing method for a specific application, such as cost, geometric complexity, material usage and properties, time, energy consumption, and sustainability. More details are reported in a previous review by the authors [9], where the classification of the most important AM techniques with a description of respective processing is reported. The advantages and disadvantages of these technologies are also reported. One of the most common problems lies with the instability of these devices, creating imperfections in the 3D printed models and differences in surface roughness across 3D reproductions with the same digital input [2]. SM techniques produce machined components with elevated accuracy and low geometrical complexity. AM allows for the creation of very complicated geometry with a lower tolerance and relative quality [10]. Several authors have proposed a solid analysis of the similarities, differences, advantages, and disadvantages of AM vs. SM through documented international works, based on these key factors. As an example, the Wohler's 2013 Report presents a wide overview of AM's incorporation into the industrial market [11], evidencing where AM can dominate SM, or otherwise cannot compete with conventional techniques. Referring to high-volume production, the high venture capital necessary to produce 3D models using AM does not make it a financially reasonable choice for producers [12]. Traditional techniques, such as, for example, injection molding, still appear more economical and suitable for this context, even if AM allows for the production of models with high geometrical complexity, removing, shaping, and linking the process by printing complete parts in a single step. An example of this is the metal acetabular cup [13] used in hip replacement surgeries; in the past it was produced using several SM and formative manufacturing processes, such as forging, machining, and coating. Contrastingly, an AM powder-bed fusion metal printer can build the acetabular cup and add the porous surface into the surface layers in a single print.

Another important field of application is mass customization, identified by its capability to offer exclusively designed products and services to each consumer because of elevated process agility, flexibility, and integration [14–16]. The lack of lead time and fast design modification in AM, together with exclusive representation supplied by 3D scanning, puts AM techniques ahead of SM processes for this specific application. Referring to the low-volume manufacture of products, which is defined specific to the industry, product, and sale capacity [17], AM manufacturers can achieve a superior complexity with equivalent costs. The launch of AM into the manufacturing sector started, in fact, with low-volume production for rapid prototyping with stereolithography. Until now, functional assembly manufacturing remains the main application of AM, as also reported by Wohler [11]. Another paper [17] compared the production cost of a small plastic bar made using AM powder-bed fusion with that of one made with injection molding, suggesting that for a production volume smaller than 10,000, AM had a reduced unit cost compared to injection molding [17]. While SM dominates the mass manufacturing region financially, AM is better suited to producing the tooling and fixtures necessary for conventional mass manufacturing molds [18,19]. AM presents a decreased lead time and cost, taking advantage of high value, low production of parts such as those used in ships, automation, aviation, and satellites, etc. AM also offers producers the capability to trade complexity for customizability in low volume production, where AM is being utilized for personally customized restoration components [20]. Nevertheless, AM machines offer production flexibility, and they are still significantly expensive in comparison to SM techniques. On the other hand, AM allows for a reduced production of wasted material and higher resource efficiency, which, importantly, has a good effect on the environmental impact, in comparison to the SM machines [4,5,9,21]. Strategic conditions for the diffusion of AM into the broader trade market involve high process stability, a database including AM material properties, on-line quality control processes, constant certification, and preparation of design guidelines. Nowadays, the growth of gradually more precise technologies, the reduction of costs, and the development of innovative high-tech materials allow for the successful use of AM techniques in several application fields, such as cultural heritage [22]. In a previous work [22], the authors demonstrated the suitability of the fused filament fabrication (FFF) method for the reproduction and restoration of four missing columns of a 17th-century polychrome wooden tabernacle belonging to a private collection. In this paper, the same columns were also reproduced with a traditional SM technique, with the aim of comparing the two different technologies in the same specific application.

2. The Case Study

This research aims to evaluate the general effectiveness of the two different methodologies in the restoration and conservation of artistic heritage; specifically, it assesses the advantages and disadvantages of each selected technique (fused filament fabrication (FFF) and computer numerical control (CNC) milling) in the reproduction of missing columns of an ancient 17th-century polychrome wooden tabernacle (Figure 1). The tabernacle is made of pine wood and according to oral sources it was handed down. With preliminary observation under the optical microscope [23], it can be seen to be decorated with an artistic technique called "Estofado de Oro".

Figure 1. 17th-century polychrome wooden tabernacle and original column.

The experimental work carried out in this paper started with the characterization of the materials used in both CNC and FFF techniques, and the subsequent comparison of some of the properties that were most interesting for conservation purposes. Next, the ancient column was replicated by using PETG for FFF and pine wood for CNC. A deep comparison analysis of the times, costs, and waste production of the two techniques was also performed before the final restoration step.

The experimental activity consisted of the following steps, summarized in Figure 2.

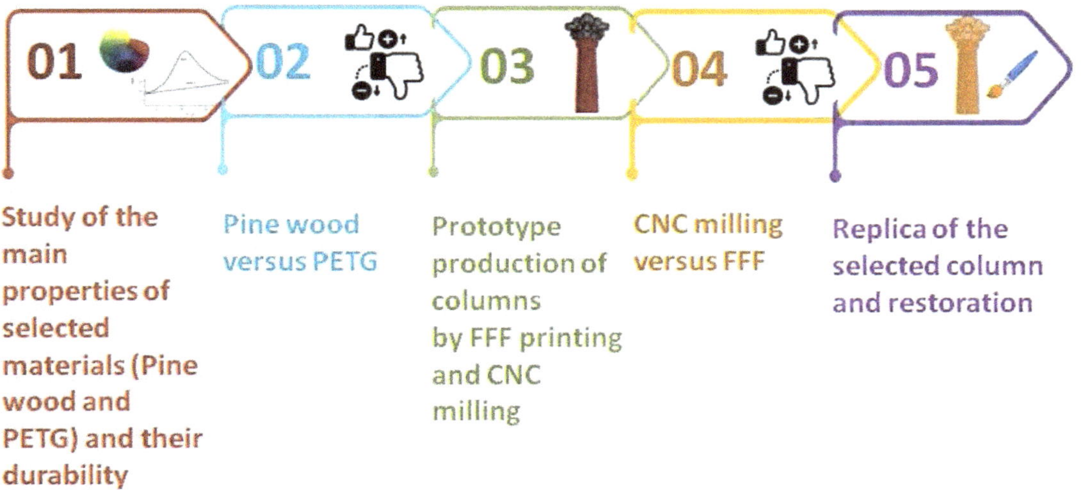

Figure 2. Experimental activity in summary.

3. Materials and Methods

3.1. Characterization of the Materials for FFF and CNC Milling Process

3.1.1. Chemical Characterization of Neat Materials

Pine wood was selected to produce the column prototype using the CNC milling technique because it is similar to the original material of the tabernacle, as reported from oral sources and confirmed by preliminary optical microscope observations. The European pine wood used in this work was supplied by Tecno Wood SRL (Lecce, Italy) with the following features: light yellow colour, straight grain, medium texture, weight 2373.79 g, length 799 mm, and square cross-section with 79 mm side.

Polyethylene terephthalate glycol (PETG), supplied by the company PrimaSELECT (Malmo, Sweden), was used for the FFF printing of the columns. It was selected by comparing its properties (such as durability and thermal stability) to that of other materials usually used in 3D printing [22]. The PETG filament has a diameter of 1.75 $\pm$ 0.05 mm, a density of 1.27 g cm^{-3}, and a melt flow index (MFI) of 12.1 g/10 min at a temperature of 225 °C, according to the supplier's data sheet.

Fourier-transform infrared spectroscopy (FTIR) analyses were performed on the neat materials (pine wood and PETG) to carry out a preliminary chemical structural characterization. FTIR spectra were obtained on KBr pellets (1 mg of wood powder and PETG and 200 mg of KBr) using a JASCO FT/IR 6300 spectrometer (Easton, MD) with a resolution of 4 cm^{-1}, setting up 64 scans in the region between 4000 and 600 cm^{-1}. Five spectra were considered for each replicate sample.

3.1.2. Durability Analysis

To study the durability of the materials used for the reproduction of the tabernacle columns, and to obtain more information on the average lifetime of the replicas, PETG and pine wood were subjected to accelerated aging tests.

The samples used for the accelerated aging test (bars with dimensions of 60 mm $\times$ 13.2 mm $\times$ 0.7 mm) were produced from pine wood and PETG, according to ISO178:2014. Specifically, the pine wood samples were produced by cutting the wooden stock piece manually, while the PETG samples were produced using the BQ HEPHESTOS 2 printer (BQ Company, Madrid, Spain) and setting the following operating conditions: extrusion temperature 220 °C, printing speed 50 mm/s, infill 20%. The CAD model was created with

Fusion 360 software (Autodesk, San Rafael, CA, USA), converted to an STL file with Cura software (Ultimaker B.V., Utrecht, The Netherlands).

Artificial aging was carried out by placing the samples in Binder model KMF115 climatic chamber (BINDER GmbH, Tuttlingen, Germany) (T = 60 ± 2 °C, RH% = 80 ± 1%) and under a xenon-arc UV lamp for 5 days [24].

Morphological structural, thermal, and mechanical analyses were carried out on test samples, before and after exposure to artificial aging, to study the durability of the materials:

- Dynamic contact angle measurements were carried out with the First Ten Angstroms FTA1000 Quick Start instrument (Newark, CA, USA) equipped with a video camera, and analyses were performed at room temperature using the sessile drop technique, according to NORMAL-33/89.
- Evaluation of color change of materials before and after artificial aging was carried out with a Konica Minolta CR-410 (Milano, Italy), equipped with a Xenon lamp. Measurements were made following the recommendations of NORMAL-43/93 and using the CIELab International Chromatic System (1976). The color changes were evaluated by the L*a*b* system and expressed as ΔE.
- XRD measurements were carried out with a Rigaku Ultima+ diffractometer (Tokyo, Japan) with CuKα radiation (λ = 1.5418 Å) in the step scan mode recorded in the 2θ range of 5–60°, with a step size of 0.02° and a step duration of 0.5 s. For each replicate sample, three spectra were considered.
- DSC analysis (Mettler Toledo DSC1 StareSystem, Milano, Italy) was performed on 3D-printed PETG samples to investigate glass transition temperature (T_g) variations, over a temperature range of 25 °C to 200 °C (heating rate of 10 °C/min).
- Flexural tests were performed on the samples using a Lloyd LR5K dynamometer (Lloyd Instruments Ltd., Bognor Regis, UK), with a test speed of 2 mm/min and a specimen support spacing of 64 mm, according to the ISO178(2014). For each sample, five replicates were made.

3.2. *Replica of the Column by FFF Printing and CNC Milling*

The CNC milling machine model HURCO VM 10 (Gindumac GmbH, Kaiserslautern, Germany) and the BQ HEPHESTOS 2 printer (BQ Company, Madrid, Spain) were used to reproduce the columns of the ancient tabernacle and to compare CNC milling versus FFF printing.

Specifically, the creation of the column replica with both methodologies followed the steps below (Figure 3):

- Morphological survey and manual drawing of the column (Figure 3A) [22];
- Transformation of the 2D model into CAD model using Rhinoceros software (Robert McNeel & Associates, Seattle, DC, USA), Figure 3B,C [22];
- Modification of the CAD model using Fusion 360 software (Autodesk, San Rafael, CA, USA); specifically, while a single CAD file was sufficient for the CNC milling technique (Figure 3D), for the 3D printing of the column it was necessary to separate the half-height 3D model into two parts and create two separate CAD files (base and capital, Figure 3E,F), due to the small print volume (210 mm × 297 mm × 220 mm) compared to the total height of the column (H 38.7 cm).
- Transformation of CAD files into G-Code and STL, using Cura software (Ultimaker B.V., Utrecht, The Netherlands);
- FFF printing of the PETG column using the following operating parameters of the 3DPRN LAB printer: 0.4 mm nozzle, 20% infill, extrusion temperature of 220 °C, platen temperature of 50 °C, printing speed of 60 mm/s, layer height of 0.2 mm;
- Reproduction of the pine wood column by CNC milling by setting the following operating parameters in the Fusion 360 software: HURCO VM 10 machine, pine wood material. Functions and cutters were set according to the different steps, as will be reported in the Section 4.2.

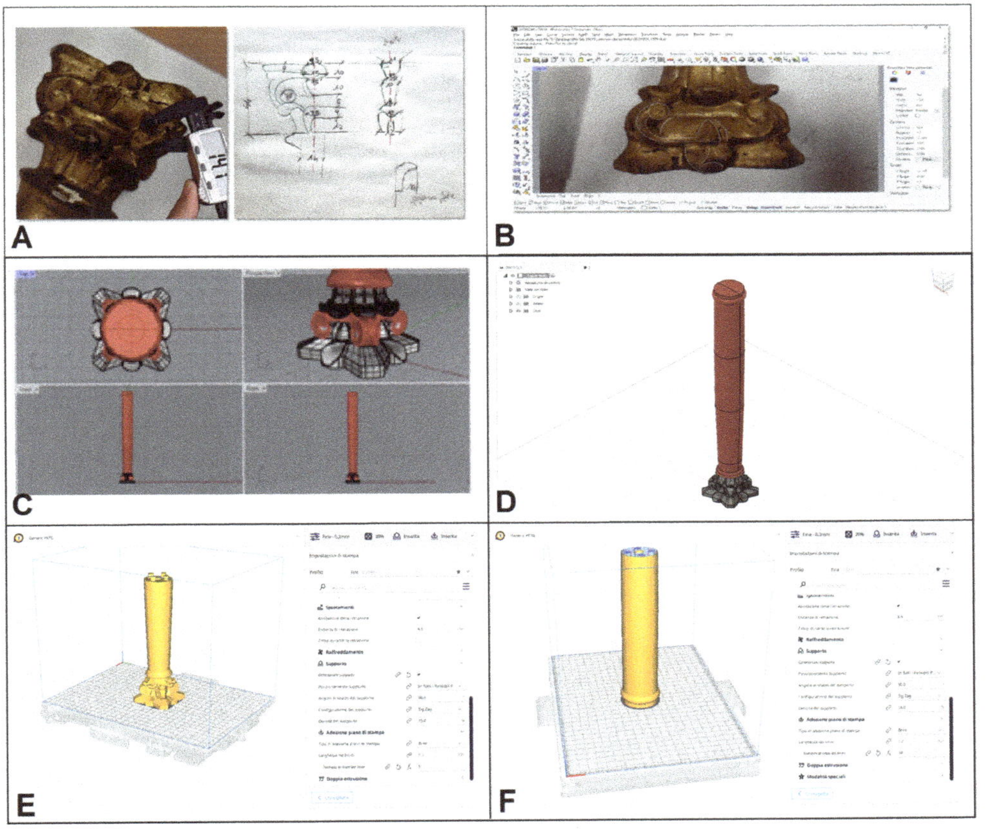

Figure 3. Surveying measurements with a caliper (left) and drawing by hand (right) (**A**); 2D drawing with Rhinoceros software (**B**); 3D modeling of the column using Rhinoceros software (**C**); CAD model modified with Fusion 360 software for CNC milling (**D**); CAD models modified with Cura software for 3D printing (**E**,**F**).

3.3. Materials for Restoration

The materials used for the restoration of columns produced were: Bologna gypsum (CTS SRL, Bari, Italy), rabbit glue (CTS SRL, Bari, Italy), putty (modostuc Gimod SRL, Pavia, Italy), yellow ochre bolus (CTS SRL, Bari, Italy), gelatin sheets (Dolciaria Pezzella SRL, Naples, Italy), 22 K gold leaf, 917/1000, dim. 8 × 8 cm (Aurum SAS gilding products, Bologna, Italy).

4. Results

4.1. Study and Characterization of Materials

The chemical characterization of the neat materials used, pine wood and PETG, was carried out by identifying the main molecular groups using FTIR spectroscopy (Figure 4). The FTIR spectra of wood samples show infrared bands at 3440, 2900, and 2800 cm^{-1} associated with O-H and C-H groups; characteristic peaks in the fingerprint region at 1509 cm^{-1} (C-C stretching of the aromatic ring) and at 1209 cm^{-1} (C-O stretching) are associated with lignin, while at 1741 cm^{-1} (C-O stretching), 1640 cm^{-1} (C-O group), and 1090 cm^{-1} (C-O stretching) are associated with polysaccharides (hemicellulose and celluloses) [25–27]. The FTIR spectra of PETG show infrared peaks at 3432 cm^{-1} (O-H group), 2925 and 2851 cm^{-1} (C-H stretching), 1743 cm^{-1} (C-O ester group), 1462 cm^{-1} (C-C stretching); 1162 cm^{-1} (ester groups), 1091 cm^{-1} (C-H bends) [28]. The spectroscopic data agree with the literature [25,26,28]. In pine wood, no peaks associated with additives, fats, or impregnants are visible, but only the bands related to the main wood components: cellulose, hemicellulose, and lignin. Furthermore, the bands at 3400 cm^{-1} related to the stretching vibrations of the hydroxyl groups are different in the FTIR spectra of the two materials [29]. It is known in the literature that the position and shape of the OH stretching band can change in the specimens, and this will be positioned at about 3350 cm^{-1} for water chemically bound to the material and 3290 cm^{-1} for water absorbed from the external environment [29]. The size of the bands in the two samples results from this phenomenon, and the fact that the band at 3400 cm^{-1} in pine wood is greater than that in PETG indicates its greater hygroscopicity.

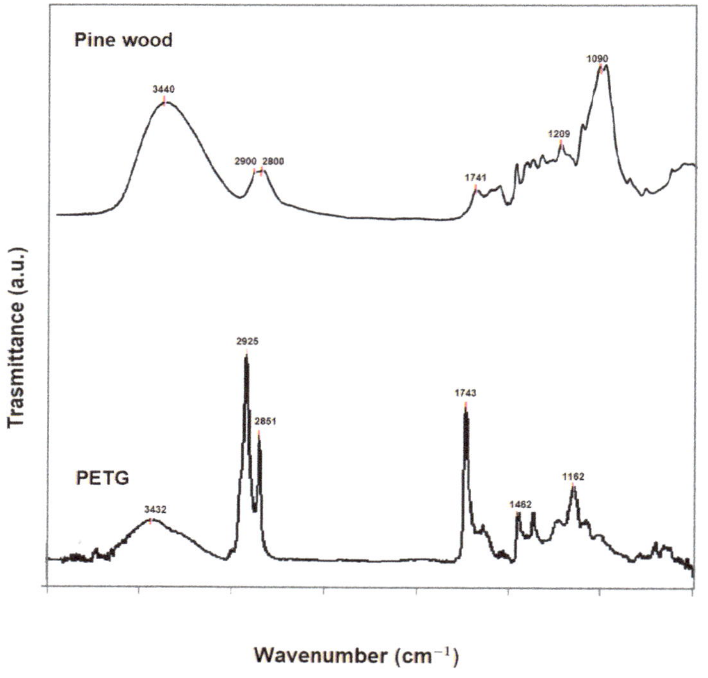

Figure 4. FTIR spectra of pine wood and PETG, and main infrared bands.

Pine wood and PETG test samples, produced as previously described, were subjected to scientific analysis before and after artificial aging in a climate chamber in order to compare the behavior of the two materials and their durability. The average results and relative standard deviations for each material are shown in Table 1.

Table 1. Comparison of surface, thermal, and mechanical properties of pine wood and PETG before and after artificial aging.

Sample	W (g)		CA (°)		ΔE		Tg (°C)		E (GPa)		ε (%)		σ (MPa)	
	Before	After	Before	After	Before	After	Before	After	Before	After	Before	After	Before	After
Pine wood	2.10 ± 0.90	2.04 ± 0.07	/	/	/	9.03	/	/	2.22 ± 0.20	0.99 ± 0.13	3.71 ± 0.03	4.11 ± 0.02	53.01 ± 2.40	40.37 ± 1.89
PETG	3.92 ± 2.11	3.90 ± 1.08	98.36 ± 3.21	95.96 ± 5.56	/	2.60	70.81 ± 0.51	71.37 ± 0.73	2.35 ± 0.02	1.13 ± 0.01	3.61 ± 0.53	3.20 ± 0.64	67.10 ± 1.20	60.20 ± 2.11

The artificial aging process had no significant effect on the size of the inspected samples. Instead, very small variations in weight were detected (Table 1). PETG preserves a good water repellency after artificial aging [30]; the contact angle value after aging is in fact 95.96 ± 5.56°, higher than the 90° limit conventionally used to distinguish a hydrophobic from a hydrophilic material [30]. In addition, it does not undergo appreciable color variations. These results agree with the literature, which reveals only a slight difference in brightness [9,24,31]. In contrast, pine wood shows a greater variation in the colorimetric coordinates L*, a*, b*, with a ΔE greater than five (Table 1). In fact, it is well known that wood is susceptible to photochemical degradation caused by exposure to light radiation, mainly due to lignin being a component [32]. Greater changes in thermal and mechanical properties are evident in the inspected samples (Figure 5). In the X-ray diffractogram of the wood sample (Figure 5A), the diffraction peaks at 15.3° and 22.2°, together with an amorphous background band, are evident; they originate from the crystalline and amorphous regions of the cellulose. Specifically, the first peak is assigned to the crystalline plane (101), while the second peak is assigned to the crystalline plane (002) [4]. The effect of artificial aging on the X-ray diffractograms of pine wood can be observed from the peak height at 22.4°, which increased significantly, in contrast to the peak height at 15.5° remaining unchanged (Figure 5A). These data are indicative of an increase in the crystallinity of the samples, as aging reduces the amorphous fractions of the wood [26]. The X-ray diffractogram of PETG (Figure 5B) shows the presence of peaks at 17.9°, 22.8°, and 26.1°, corresponding to crystal planes with Miller indexes of (010), (110), and (100), respectively [33]. As a result of artificial aging, an increase in the amorphous phase of PETG is observed, with a consequent decrease in the crystalline phase (Figure 5B). PETG neat shows a glass transition temperature (T_g) of about 70.8 ± 0.5 °C (Figure 5C,D), calculated using the inflection point method and comparable with the reference literature [22,28]. After aging in a climate chamber, T_g of PETG is slightly reduced, indicating a decrease in the degree of crystallinity (Table 1, Figure 5C,D), which was previously indicated by XRD analysis. Accelerated aging can usually cause depolymerization of polymers, resulting in a less ordered structure [34]. This phenomenon involves the decomposition of the polymer into low molecular weight molecules, which occurs spontaneously through a natural aging process or, as in our artificial study, due to the influence of temperature and humidity [34,35]. Mechanical test data show different bending behaviors in the two materials (Figure 5E,F). This difference is accentuated by artificial aging and is more evident in pine wood. In particular, while the flexural modulus of elasticity (GPa) and the flexural stress (MPa) of both materials decrease after aging, contrastingly flexural strain (%) increases in pine wood (Figure 5E).

A general comparison of the morphological structural features of the two materials subjected to artificial aging shows a higher durability of PETG compared to pine wood, as was also reported in the literature [22,24,32,34,36]. However, to be exhaustive, the comparison must be applied to the specific case study analyzed and, thus, to the reproduction of the same column. In fact, although both materials are subject to a decrease in crystallinity and mechanical performance with aging, these characteristics are less evident in PETG, which preserves the water-repellent and colorimetric properties that are of fundamental importance to the preservation of the structural and chromatic properties of the columns and tabernacle. Furthermore, the increase in bending deformation manifested by pine wood with aging could cause greater instances of discontinuity surfaces in the multilayer system of the column (pictorial layers/gold leaf/stucco/support) and, consequently, more problems over time due to the detachment of the pictorial layers and the formation of lacunae on the column. As a result, PETG's specific properties of good resistance and stability are more suitable for column reproduction for conservation purposes.

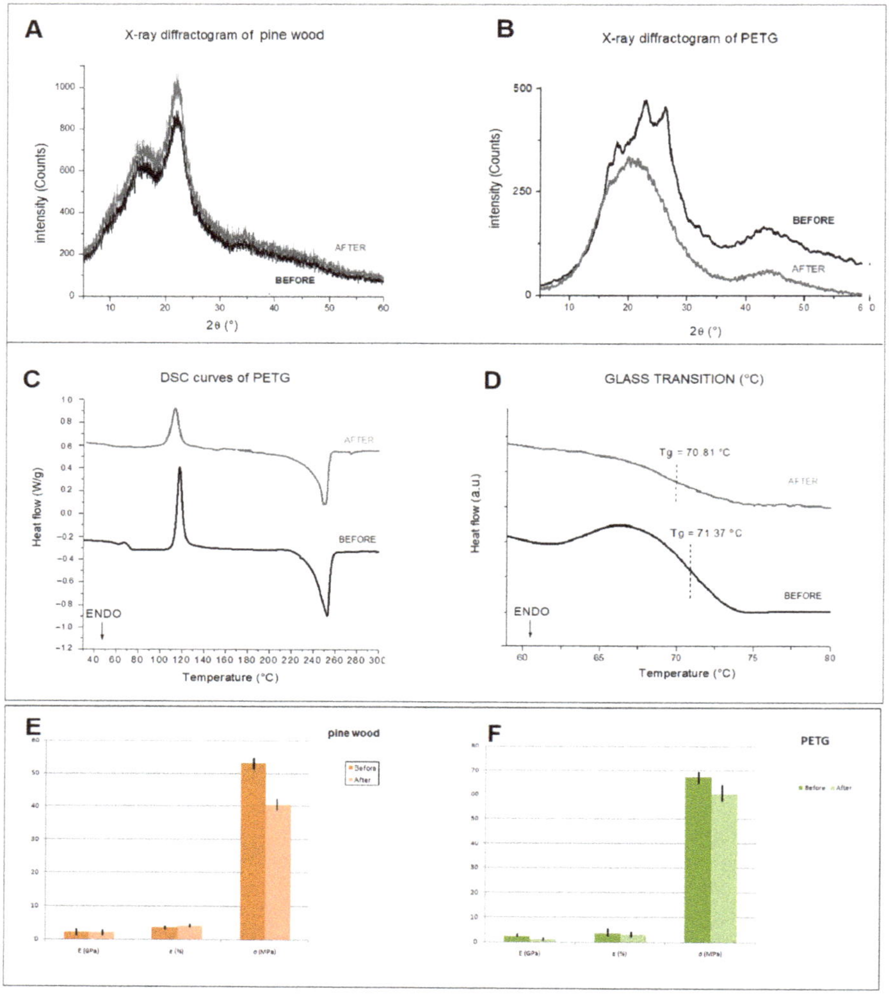

Figure 5. X-ray diffractogram of pine wood (**A**) and PETG (**B**) before and after aging; DSC curves of PETG (**C**) and glass transition temperature (**D**) before and after aging; flexural properties of pine wood (**E**) and PETG (**F**) before and after aging.

4.2. Replica of the Column by FFF Printing and CNC Milling

A replica of the column was reproduced in PETG using the FFF technique (Figure 6A,C). Specifically, the printing strategy used facilitated the positioning of the object on the printer's plate, eliminating the use of supports (thus with less waste of raw material). Furthermore, it allowed for layers perpendicular to the height of the column, enabling, during the final restoration phase, a good grip on the layers applied by the restorer to recreate the original texture. The following process parameters were used: nozzle 0.4 mm, infill 20%, extrusion temperature of 220 °C, plate temperature 50 °C, printing speed 60 mm/s, and layer height 0.2 mm. All details of the 3D printing can be found in our previous work [22].

For the reproduction in pine wood of the column by means of CNC milling (Figure 6B,C), the machine was first initialized (including warm-up and calibration phases), and the tools were assembled and set up (selection of tool position, assembly of the spindle tool holder collet, assignment of the identification number and probing cycle). Subsequent operations involved the assembly of the blank on the four-jaw chuck (considering a length of the blank greater than the actual length of the workpiece) and the setup of the workpiece zero, the position of which must coincide with the choice made during the design phase on the CAM software (Fusion 360 software (Autodesk, San Rafael, CA, USA)). Once the CNC machine setup was complete, the G-code file was uploaded. Due to the non-uniformity of the surface and section of the wood piece used, an initial roughing/finishing of the workpiece was carried out using the "parallel" function and a Ø 20 mm milling cutter (Figure 7A). The passes were carried out parallel to the plane selected, and the tool path followed the surface in the Z direction. Then, the workpiece was blocked in the chuck, placing the jaws on the processed area. The processing was continued by always using the "parallel" function and the following parameters: a Z step of 10 mm, a side step of 8 mm, a cutting feed rate of 1300 mm/min, a spindle speed of 5000 rpm, and an overmetal value of 5 mm. The milling cutter used was similar to the previous process. By contrast, the second processing was undertaken using the "parallel" function by a single pass, a side step of 6 mm, and the value of the supermetal reduced to 1 mm (Figure 7A). Finally, to reduce the average roughness, the third processing was carried out using a Ø 5 mm milling cutter, a lateral step value of 1 mm, and an oversize of 0.5 mm. After finishing the roughing operations, workpiece finishing was carried out. Initially, the capital was processed with a Ø 5 mm and a 0.25 mm lateral step milling cutter, obtaining an average ridge height value of 0.032 mm. Subsequently, the finishing of the capital was completed by using a Ø 1 mm cutter and a lateral step value of 0.1 mm (Figure 7B). The last operations involved finishing the truncated conical lower part of the column (Figure 7C). In this case, the "linear rotary multi-axis" function was chosen (lateral step equal to 1 degree and cutter Ø 5 mm). Partial truncations (90 degrees apart) were made using the "2D contouring" function (three passes with a Z step of 5 mm) and, finally, a manual truncation of the finished product.

Figure 6. PETG column printing with BQ HEPHESTOS 2 printer (**A**); pine wood column production with HURCO VM 10 (**B**); comparison between the two columns produced (**C**).

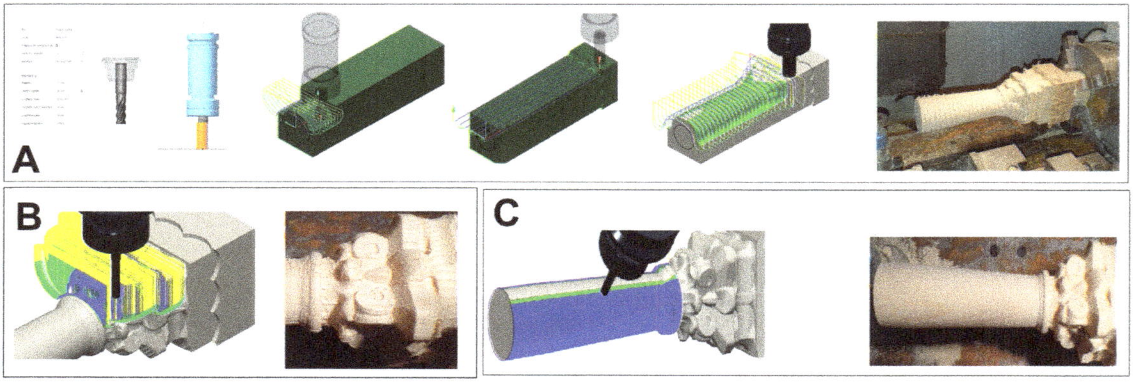

Figure 7. Initial roughing of the workpiece with the "parallel" function and a Ø 20 mm milling cutter (**A**); finishing operations on the capital (**B**); finishing operations on the lower truncated conical part of the column with the "rotary linear multi-axis" function (**C**).

4.3. CNC Milling versus FFF: The Case Study of the Missing Column of the Tabernacle

A general comparison between the advantages and disadvantages of the two analytical techniques investigated, CNC milling and FFF printing, is given as Supplementary Material S1.

In this section, the authors examine the same parameters for each technique, analyzing them according to the specific application: the reproduction of the missing columns of the 17th-century polychrome wooden tabernacle. There are, in fact, a lot of overlaps concerning the range of different possible applications. It is well known in the literature that CNC milling represents the best manufacturing solution for producing strong, accurate, and heat-resistant objects. However, FFF could be more suitable for different application fields. In this paper, we focus on a cultural heritage conservation application. We compare the different design and process steps required by both CNC milling and a low-cost FFF machine to produce the same column of a 17th-century polychrome wooden tabernacle, evidencing the advantages and disadvantages of the two building methodologies used for the same application. It is important to underline that the same column was printed using the FFF method with PETG and in the CNC milling machine, by using the same wood that the original tabernacle was made from.

The comparison of the properties of the two materials and the analysis of the accuracy of the models produced have been reported in previous Sections 4.1 and 4.2. Here, the construction time, material waste, and total production cost in both cases were calculated and summarized in the following tables.

The time required for each building step necessitated by the two different technologies was determined by the laboratory technician using the information provided by the machines' software, and it was also recorded using the stopwatch. The results obtained are reported in Table 2.

The calculation reported in Table 2 suggests that, for this specific application, CNC milling is more rapid than FFF according to the general considerations, as previously reported.

Table 3 reports a comparison of the costs required by the two building methodologies.

Specifically, Table 3 shows the material cost, operator cost, and tooling cost for both FFF and CNC milling. The material costs were determined by using the costs of materials reported on the technical data sheet, considering the amount of each material required by both FFF and CNC milling techniques to produce the same object. In particular, 210 g of PETG and 2400 g of pine wood were required by FFF and CNC milling, respectively. The operator cost for both the CNC and FFF was given by the lab technician, considering the unit cost of the operator and the total time reported in Table 2, without the duration of the building process (i.e., 15 h for FFF and 5 h for CNC milling). The tooling cost of each machine was determined by considering the price of each machine. In particular, the cost of FFF is EUR 1000, while the cost of CNC milling is EUR 11,0000. As a consequence of the data reported in Table 3, it is evident that the total production costs of CNC milling are a great deal larger than those of FFF. Finally, the mass of wood and PETG waste produced during the CNC milling and FFF process were calculated. In the case of FFF, the mass of waste is negligible since it was not necessary to produce supports, and, hence, no waste materials were obtained at the end of the 3D printing process. The calculation of wood waste mass (measured in grams) produced by the CNC milling process is reported as follows:

$$\text{Wood waste mass} = \text{mass of the neat wood stock} - \text{mass of the produced column}$$

The mass of the neat wood stock was measured by weighing the piece before using it; it is equal to 2374 g. The mass of the produced column was also determined by weighing it; it is equal to 240 g. It is, thus, evident that the CNC milling process produces a very large amount of waste material (about 2134 g).

Finally, Table 4 summarizes all the data collected by comparing the two techniques for the production of the column of the tabernacle.

Table 2. Comparison between time (h) of CNC milling and FFF for the production of a column of the 17th-century tabernacle.

Time (h)	FFF	CNC Milling
Design and Machining Strategies	2.0	4.0
Selection of Process Parameters	0.5	1.0
Simulation	1.0	1.5
Building Processing	15.0	5.0
Machine Cleaning	0	1.0
Total	18.5	12.5

Table 3. Comparison of costs of CNC milling and FFF for the production of a column of the 17th-century tabernacle.

Costs (Euro)	FFM	CNC Milling
	PETG	**Pine Wood Stock**
Material Cost	9.45	2.50
Operator Cost	375.00	125.00
Machine Cost	1000.00	110,000.00
Total	1384.45	110,127.5

Table 4. Comparison between costs of CNC milling and FFF for the production of a column of the 17th-century tabernacle.

	FFF	CNC Milling
	PETG	**Pine Wood Stock**
Time (h)	18.5	12.5
Costs (EUR)	1429	110.000
Material Wastage (g)	No wastage	2134

4.4. Restoration

Based on the results obtained from the materials' characterization and mainly from the comparison of the two production methods used (FFF and CNC milling), which showed the greater suitability of FFF as a technique to produce the missing column, the final restoration operations were carried out by the restorer exclusively on the PETG column, as reported: preparation of the substrate with Bologna plaster and rabbit glue, modeling of the stucco with a scalpel, application of red bolus, application of gold leaf with isinglass, and burnishing (Figure 8C). The success of the restoration is unmistakably shown by the images in Figure 8, clearly demonstrating the suitability of FFF techniques for this specific application, as well as the suitability of PETG for this conservative restoration.

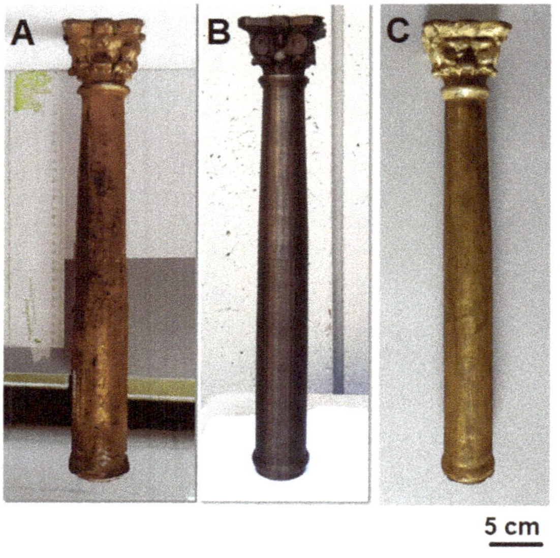

Figure 8. Original column (**A**); PETG printed column replica (**B**); and PETG column after restoration (**C**).

5. Conclusions

In this paper, one of the of four missing columns of a 17th-century polychrome wooden tabernacle was reproduced using two of the most popular subtractive and additive technologies—FFF and CNC milling—with the aim of selecting the most suitable machine for this specific application. The analysis of the durability of the different materials used in the two machines, European pine wood (for CNC milling) and PETG (for FFF) confirmed that the mechanical properties of both materials decrease after accelerated aging weathering. However, PETG's specific features of higher mechanical and wet resistance, stability and preservation of chromatic properties made it more suitable than pine wood for column reproduction for conservation purposes. This preliminary result was not a sufficient basis on which to select one of the two techniques, basing the decision only on the performances of the used material. For this reason, a deeper analysis based on a comparison between time, cost, and waste production of the two machines was performed. The calculations demonstrated that the building time of FFF is higher than that of CNC. However, the production costs and waste materials connected to the CNC process were much greater than those connected to FFF. Starting from this result, and considering that the different materials used for the realization of the two columns does not positively affect the steps of restoration, the authors selected the FFF process as the most suitable technique for producing the missing column. The 3D printed PETG column was, hence, successfully restored, clearly demonstrating the suitability of 3D printing for use in cultural heritage applications.

Supplementary Materials: The following supporting information can be downloaded at: https://www.mdpi.com/article/10.3390/ma16083038/s1, Supplementary Materials S1: CNC milling versus FFF. References [1,9] are cited in the supplementary materials.

Author Contributions: Conceptualization, D.F., D.R. and C.E.C.; software, F.M., D.R. and D.F.; writing—original draft preparation, D.F. and C.E.C.; writing—review and editing D.F., F.M., D.R. and C.E.C.; supervision, C.E.C. All authors have read and agreed to the published version of the manuscript.

Funding: This research received no external funding.

Institutional Review Board Statement: Not applicable.

Informed Consent Statement: Not applicable.

Data Availability Statement: Not applicable.

Acknowledgments: The authors thank Valentina De Carolis, Fulvio Cortese, Paride Fedele for the support provided in the 3D modeling and Nicola Ancona for the restoration.

Conflicts of Interest: The authors declare no conflict of interest.

References

1. Pereira, T.; Kennedy, J.V.; Potgieter, J. A Comparison of Traditional Manufacturing vs Additive Manufacturing, the Best Method for the Job. *Procedia Manuf.* **2019**, *30*, 11–18. [CrossRef]
2. Gao, W.; Zhang, Y.; Ramanujan, D.; Ramani, K.; Chen, Y.; Williams, C.B.; Wang, C.C.L.; Shin, Y.C.; Zhang, S.; Zavattieri, P.D. The Status, Challenges, and Future of Additive Manufacturing in Engineering. *CAD Comput. Aided Des.* **2015**, *69*, 65–89. [CrossRef]
3. Esposito Corcione, C.; Gervaso, F.; Scalera, F.; Padmanabhan, S.K.; Madaghiele, M.; Montagna, F.; Sannino, A.; Licciulli, A.; Maffezzoli, A. Highly Loaded Hydroxyapatite Microsphere/ PLA Porous Scaffolds Obtained by Fused Deposition Modelling. *Ceram. Int.* **2019**, *45*, 2803–2810. [CrossRef]
4. Fico, D.; Rizzo, D.; De Carolis, V.; Montagna, F.; Palumbo, E.; Corcione, C.E. Development and Characterization of Sustainable PLA/Olive Wood Waste Composites for Rehabilitation Applications Using Fused Filament Fabrication (FFF). *J. Build. Eng.* **2022**, *56*, 104673. [CrossRef]
5. Esposito Corcione, C.; Palumbo, E.; Masciullo, A.; Montagna, F.; Torricelli, M.C. Fused Deposition Modeling (FDM): An Innovative Technique Aimed at Reusing Lecce Stone Waste for Industrial Design and Building Applications. *Constr. Build. Mater.* **2018**, *158*, 276–284. [CrossRef]
6. Corcione, C.E.; Gervaso, F.; Scalera, F.; Montagna, F.; Maiullaro, T.; Sannino, A.; Maffezzoli, A. 3D Printing of Hydroxyapatite Polymer-Based Composites for Bone Tissue Engineering. *J. Polym. Eng.* **2017**, *37*, 741–746. [CrossRef]
7. Romani, A.; Rognoli, V.; Levi, M. Design, Materials, and Extrusion-Based Additive Manufacturing in Circular Economy Contexts: From Waste to New Products. *Sustainability* **2021**, *13*, 7269. [CrossRef]
8. Mantelli, A.; Romani, A.; Suriano, R.; Diani, M.; Colledani, M.; Sarlin, E.; Turri, S.; Levi, M. Uv-Assisted 3d Printing of Polymer Composites from Thermally and Mechanically Recycled Carbon Fibers. *Polymers* **2021**, *13*, 726. [CrossRef]
9. Fico, D.; Rizzo, D.; Casciaro, R.; Esposito Corcione, C. A Review of Polymer-Based Materials for Fused Filament Fabrication (FFF): Focus on Sustainability and Recycled Materials. *Polymers* **2022**, *14*, 465. [CrossRef]
10. Newman, S.T.; Zhu, Z.; Dhokia, V.; Shokrani, A. Process Planning for Additive and Subtractive Manufacturing Technologies. *CIRP Ann. Manuf. Technol.* **2015**, *64*, 467–470. [CrossRef]
11. Wholers, T. *Wohler's Report 2013*; Wholers Associates: Washington, DC, USA, 2013; ISBN 0-9754429-9-6.
12. Kalpakjian, S.; Schmid, S.R.; Sekar, K.S.V. *Manufacturing Engineering and Technology*; Pearson Education South Asia: Singapore, 2014; ISBN 9789810694067.
13. Unger, A.S.; Lewis, R.J.; Gruen, T. Evaluation of a Porous Tantalum Uncemented Acetabular Cup in Revision Total Hip Arthroplasty: Clinical and Radiological Results of 60 Hips. *J. Arthroplast.* **2005**, *20*, 1002–1009. [CrossRef]
14. Pine, I.; Joseph, B.; Victor, B. *Making Mass Customization Work*; Harvard Business Publishing: Harvard, MA, USA, 1993.
15. Eastwood, M.A. Implementing Mass Customization. *Comput. Ind.* **1996**, *30*, 171–174. [CrossRef]
16. Tseng, M.M.; Jiao, J.; Merchant, M.E. Design for Mass Customization. *CIRP Ann. Manuf. Technol.* **1996**, *45*, 153–156. [CrossRef]
17. Conner, B.P.; Manogharan, G.P.; Martof, A.N.; Rodomsky, L.M.; Rodomsky, C.M.; Jordan, D.C.; Limperos, J.W. Making Sense of 3-D Printing: Creating a Map of Additive Manufacturing Products and Services. *Addit. Manuf.* **2014**, *1*, 64–76. [CrossRef]
18. Upadhyay, M.; Sivarupan, T.; El Mansori, M. 3D Printing for Rapid Sand Casting—A Review. *J. Manuf. Process.* **2017**, *29*, 211–220. [CrossRef]
19. Levy, G.N.; Schindel, R.; Kruth, J.P. Rapid Manufacturing and Rapid Tooling with Layer Manufacturing (LM) Technologies, State of the Art and Future Perspectives. *CIRP Ann. Manuf. Technol.* **2003**, *52*, 589–609. [CrossRef]
20. Dutta, B.; Froes, F.H. *Additive Manufacturing of Titanium Alloys Additive Manufacturing of Titanium Alloys State of the Art, Challenges, and Opportunities*; Butterworth-Heinemann: Oxford, UK, 2014; ISBN 9780128047828.
21. Fico, D.; Rizzo, D.; De Carolis, V.; Montagna, F.; Esposito Corcione, C. Sustainable Polymer Composites Manufacturing through 3D Printing Technologies by Using Recycled Polymer and Filler. *Polymers* **2022**, *14*, 3756. [CrossRef]

22. Rizzo, D.; Fico, D.; Montagna, F.; Casciaro, R.; Esposito Corcione, C. From Virtual Reconstruction to Additive Manufacturing: Application of Advanced Technologies for the Integration of a 17th-Century Wooden Ciborium. *Materials* **2023**, *16*, 1424. [CrossRef]
23. Fico, D.; Faraco, M.; Pennetta, A.; Rizzo, D.; De Benedetto, G.E. Characterization of Materials and Artistic Techniques on Two 17th-Century Neapolitan Wood Sculptures. *Archaeometry* **2018**, *60*, 834–844. [CrossRef]
24. Higueras, M. Colorimetric Evaluation of 3D Printing Polymers Exposed to Accelerated Aging for Cultural Heritage Applications. *Color Res. Appl.* **2023**, 1–13. [CrossRef]
25. Traoré, M.; Kaal, J.; Martínez Cortizas, A. Differentiation between Pine Woods According to Species and Growing Location Using FTIR-ATR. *Wood Sci. Technol.* **2018**, *52*, 487–504. [CrossRef] [PubMed]
26. Lionetto, F.; Del Sole, R.; Cannoletta, D.; Vasapollo, G.; Maffezzoli, A. Monitoring Wood Degradation during Weathering by Cellulose Crystallinity. *Materials* **2012**, *5*, 1910–1922. [CrossRef]
27. Lettieri, M.; Giannotta, M.T. Investigations by FT-IR Spectroscopy on Residues in Pottery Cosmetic Vases from Archaeological Sites in the Mediterranean Basin. *Int. J. Exp. Spectrosc. Tech.* **2017**, *2*, 2–10. [CrossRef] [PubMed]
28. Latko-Durałek, P.; Dydek, K.; Boczkowska, A. Thermal, Rheological and Mechanical Properties of PETG/RPETG Blends. *J. Polym. Environ.* **2019**, *27*, 2600–2606. [CrossRef]
29. Derrick, M.R.; Stulik, D.; Landry, J.M. *Infrared Spectroscopy in Conservation Science—Scientific Tools for Conservation*; The Getty Conservation Institute: Los Angeles, CA, USA, 1999; ISBN 9788578110796.
30. Law, K.Y. Water-Surface Interactions and Definitions for Hydrophilicity, Hydrophobicity and Superhydrophobicity. *Pure Appl. Chem.* **2015**, *87*, 759–765. [CrossRef]
31. Cano-Vicent, A.; Tambuwala, M.M.; Hassan, S.S.; Barh, D.; Aljabali, A.A.A.; Birkett, M.; Arjunan, A.; Serrano-Aroca, Á. Fused Deposition Modelling: Current Status, Methodology, Applications and Future Prospects. *Addit. Manuf.* **2021**, *47*, 102378. [CrossRef]
32. Kránitz, K.; Sonderegger, W.; Bues, C.T.; Niemz, P. Effects of Aging on Wood: A Literature Review. *Wood Sci. Technol.* **2016**, *50*, 7–22. [CrossRef]
33. Ferrari, F.; Esposito Corcione, C.; Montagna, F.; Maffezzoli, A. 3D Printing of Polymer Waste for Improving People's Awareness about Marine Litter. *Polymers* **2020**, *12*, 1738. [CrossRef]
34. Shi, Q.; Xiao, R.; Yang, H.; Lei, D. Effects of Physical Aging on Thermomechanical Behaviors of Poly(Ethylene Terephthalate)-Glycol (PETG). *Polym. Technol. Mater.* **2020**, *59*, 835–846. [CrossRef]
35. Amza, C.G.; Zapciu, A.; Baciu, F.; Vasile, M.I.; Nicoara, A.I. Accelerated Aging Effect on Mechanical Properties of Common 3d-printing Polymers. *Polymers* **2021**, *13*, 4132. [CrossRef]
36. Jankowska, A.; Artur, W.; Andrzej, M. The Influence of Artificial Weathering on Changes in Color of Selected Coniferous Wood Species. *Ann. Warsaw Univ. Life Sci. SGGW. For. Wood Technol.* **2014**, *85*, 95–100.

materials MDPI

Article

The Historic Materials and Structures Due to the Aspect of Their Actual Challenges

Dariusz Bajno [1], Agnieszka Grzybowska [1] and Ireneusz Trzyński [2,*]

[1] Department of Building Structures, Faculty of Civil and Environmental Engineering and Architecture, Bydgoszcz University of Science and Technology, 85-796 Bydgoszcz, Poland
[2] Ag-Cel Laboratory, ul. Bydgoska 14, 89-620 Chojnice, Poland
* Correspondence: irek.marek@wp.pl; Tel.: +48-785-475-724

Abstract: The subject of the article is to assess the suitability of over materials over a hundred years old that are embedded in historic building structures in conditions of contemporary utility challenges in residential and public buildings. It is based on an example of technical condition evaluation of a ceramic staircase erected in 1840 and a structure of two reinforced concrete staircases from the year 1910. As a part of in-situ and laboratory tests, the physical and mechanical parameters of unique historical materials (brick, concrete) were determined. Then the conditions for their incorporation were inventoried and determined in order to save the unique material and technical solutions used in the first half of the 19th century and the first decade of the 20th century. The article is a summary of the research and analyzes carried out in terms of proper handling of historical materials, buildings and their elements that could still fulfill their original function and be a witness to the history of a certain era. Both a research case and an application case are described here. It will allow for the continuation of these studies directly in the facility, thus assessing the effectiveness and suitability of such methods for use in similar or other situations. The aim of this approach was to introduce a non-invasive reinforcing technique that would not change the valuable and authentic appearance of these historic structures. It would also not change their static schemes, and at the same time would ensure their proper load-bearing capacity, bearing in mind that the materials used here are not equivalent to current regulations and standards.

Keywords: staircases exploitations; buildings diagnostics; historic buildings; concrete/ceramic staircases; in-situ/laboratory tests; physical/mechanical parameters

Citation: Bajno, D.; Grzybowska, A.; Trzyński, I. The Historic Materials and Structures Due to the Aspect of Their Actual Challenges. *Materials* **2023**, *16*, 2302. https://doi.org/10.3390/ma16062302

Academic Editors: Francesca Ceroni and Oldrich Sucharda

Received: 7 January 2023
Revised: 11 February 2023
Accepted: 10 March 2023
Published: 13 March 2023

1. Introduction

The inspiration to take up the presented topic was the inquiry of the Provincial Office for the Protection of Monuments about the possibility of leaving and further to use the historic staircases in renovated buildings in new conditions. Designers of renovation, strengthening and modernization works had assigned these structures to demolition due to the low strength of concrete and because of currently not used technologies [1–25].

Authors of the article decided to evaluate these materials and use them in structures in terms of suitability for today's construction and operational requirements. As it was mentioned above, the subject of the research was over a hundred-year-old materials, of which the structure of three staircases marked as K1, K2 and K5 (ceramic) were made. The main idea was to leave them for further use in a new function of the building, i.e., the Centre for Vocational Education, and indicating the scope and solutions of their possible reinforcements. The research, the accurate inventory of three subjective structures and also the research of the elaboration [26–32] were used to carry out the static and strength analysis and to indicate the reinforcement zones where the materials did not show sufficient load-bearing capacity.

The ceramic (K5) and concrete (K1, K2) staircases responsible for vertical communication in buildings have been so far used as fully-fledged load-bearing elements in a public utility facility (hospital facility with clinics), until it was shutdown. An object of this rank had to be regularly inspected, but at the time of performing these elaborations [1,2], their authors did not have such documents at their disposal. Inspection of the above-mentioned structures, carried out in December 2021 and April 2022, did not find any deformations and damage to these structures in terms of excessive deflection, cracks or even scratches exceeding the limits considered acceptable, i.e., 0.3 mm. Initially, the condition of the structure was considered to be in a good technical condition and suitable for further exploitation.

In terms of diagnostics, it is also important to refer to the applicable regulations and standards. For calculating the load-bearing capacity of the staircase structure, Polish Standards were used [20–25], older than the current ones, which were considered to be closer to the historical solutions at the time when the building was erected with no obligatory standards. An example is the class of concrete. In the current standards, the lowest strength class is C12/15 (former designation B15), while at the time of the staircase construction, these were classes from the B3 to B5 range, in special cases B7 (B7.5). Such classes can still be found with the Polish Standard from year 1976 [22].

2. Materials and Methods

The aim of the work was to make the research of building materials (shown in Tables 1–3) and to determine the static schemes that were necessary to perform the calculations that allows to verify the load capacity of the staircases, e.g., in order to propose the appropriate reinforcements, where the bearing capacity has been exhausted. As a result of the analysis, a secondary aim emerged, i.e., to propose reinforcements for areas that were exhausted in terms of their bearing capacity. As another method, calculations performed on the basis of the strength results of tested materials should be indicated. Calculations were made with our own algorithms and calculation programs based on FEM, such as Autodesk Robot Structural Analysis. Conducted moisture tests of bricks and concrete of the staircases' structures showed an air-dry condition of moisture, in which the mass moisture content of the mentioned materials did not exceed 3%. No traces of moisture were found to have been caused by leaky roofs or capillary humidity rise in those staircases.

2.1. Loads

Both values of permanent and functional loads were adopted on the basis of Polish Standards [20,21], with corresponding load factors. This corresponded to the values of the loads assumed for the design of the structure during the erection period of the buildings. The usable load of staircases was assumed in accordance with "Table 2a Communication spaces" [21], with a value of 4.0 kN/m^2. It is the number of loads required to be transferred on communication areas in office buildings, schools, research institutes, banks and medical clinics. The specifics of the above-mentioned rooms do not differ much from those in hospitals and prison buildings, where the required operational loads are 1.0 kN/m^2 less, the value being 3.0 kN/m^2. This load is classified as a long-term, in fact it will be a short-term or even temporary load. In the combination of loads, only a part of it (its short-term value according to PN-82/B-02003) described by the coefficient $\psi d = 0.35$ could be assumed, but its full value was taken into account in the verification calculations due to the fact that these structures have already passed the technical service life that were assumed for them (over 100 years). The structures of all three staircases have been verified due to capacity of the load-bearing characteristic value of 4.0 kN/m^2.

2.2. The Perfomance Technologies of Staircases

2.2.1. Concrete and Iron (Steel) for Reinforcement Bars—K1 and K2 Reinforced Concrete Staircase

The first tests of concrete compressive strength [30–32], carried out on three samples, revealed a very low class of concrete, of the order of 9.2 MPa ÷ 12.3 MPa (the middle value was 10.0 MPa), which did not allow for classification within the concrete class currently in construction. Nevertheless, in first three decades of the 20th century, concrete classes with kb = 30, 35, 45, 45, 50 kg/cm^2 (currently they should be defined as B3 ÷ B5, there is no reference to Eurocode 6) were commonly used in concrete structures and concrete reinforced with iron (steel) bars [25,33]. In the first decade of the 20th century, design studies of concrete structures at that time, including stairs, class B7.5 (E_b = 140,000 kg/cm^2) concrete was also used and iron for reinforcement bars with E_e = 210,000 kg/cm^2 [33] fragment of the description below—Figure 1 (original, translated from German language).

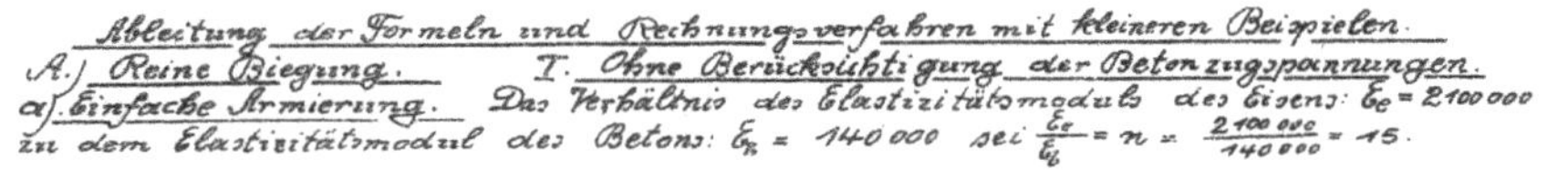
Ableitung der Formeln und Rechnungsverfahren mit kleineren Beispielen.
A.) Reine Biegung. I. Ohne Berücksichtigung der Betonzugspannungen.
a.) Einfache Armierung. Das Verhältnis des Elastizitätsmoduls des Eisens: E_e = 2100000
zu dem Elastizitätsmodul des Betons: E_b = 140 000 sei $\frac{E_e}{E_b} = n = \frac{2100000}{140000} = 15$.

Figure 1. A fragment of a German-language manuscript that describes properties of materials [33]. Translation below.

"A derivation of formulas and calculation methods with small examples

A. *Pure bending I. without considering concrete tensile stresses.*

(a) *Simple reinforcement. The ratio of the modulus of elasticity of iron ε_e = 2,100,000 kg/cm^2 to the modulus of elasticity of concrete ε_b = 140,000 kg/cm^2*

$\varepsilon_e/\varepsilon_b = n$ = 2,100,000/140,000 = 15"

The research of concrete structures from the beginning of 20th century indicated much higher strength parameters than assumed by their original designers. An example of this is the research carried out by another laboratory, which has checked the mechanical parameters of concrete and bricks by the request of the following article authors.

Basing on the compression research of 3 concrete samples according to [30], it was possible to assume the B12.5 concrete class, which was used quite often in Poland and was even used in the second half of the 20th century to design and manufacture concrete and reinforced concrete structural elements. Concrete testing for the K1 staircase (presented in the following sections), using both the non-destructive and laboratory methods, indicated much higher compressive strength classes of concrete [28]. On their basis, it was assumed that on average it will be class C20/25 (B25), but for the verification calculations, lower parameters of structural concrete, corresponding to class C12/15 (B15), were adopted. Additionally, the load-bearing capacity of these structures was checked, assuming that they were made of B12.5 class concrete. Additionally, a series of "in situ" research has been carried out using the "Schmidt's hammer" sclerometer, not simply to estimate the concrete class, but to evaluate the homogeneity in its structure.

It is similar with reinforcing steel (iron, in this case, probably cast). According to the study [25], the immediate tensile strength of steel with a modulus of elasticity E = 2,100,000 kg/cm^2 ranges from 3700 kg/cm^2 (370 MPa) to 5800 kg/cm^2 (580 MPa). According to the studies [34], the tensile strength of cast steel was in the range of Rm = 370 ÷ 450 MPa, so it was practically consistent with the source cited earlier. These steels can be considered as the equivalent of St3S steel according to PN-88/H-84020 with Rm = 360 ÷ 490 MPa. This is also confirmed by the research [31], where basing on the measurement of steel hardness, the tensile strength Rm = 235 MPa was assigned to it.

This section presents the results of research carried out for the structural elements made of reinforced concrete of the K1 staircase, which also had a reinforced concrete railing that served as the stringers of the running plates. The structural thickness of the running plate and landing is 13 cm, and the utility layer is 3 cm thick terrazzo (Figure 2a). The reinforced concrete balustrade is monolithised (composite) with the stair tread and constitutes the above-mentioned cheek beam. Figures 2a and 3 shows the place of collecting the tested samples. Representative sample No. 7 was collected in the form of a core in the staircase, which was then tested in a strength press. Sample No. 5 was taken from the landing. The test was carried out in the Pilot 4 testing machine in accordance with the applicable standards, the results are presented in Table 2.

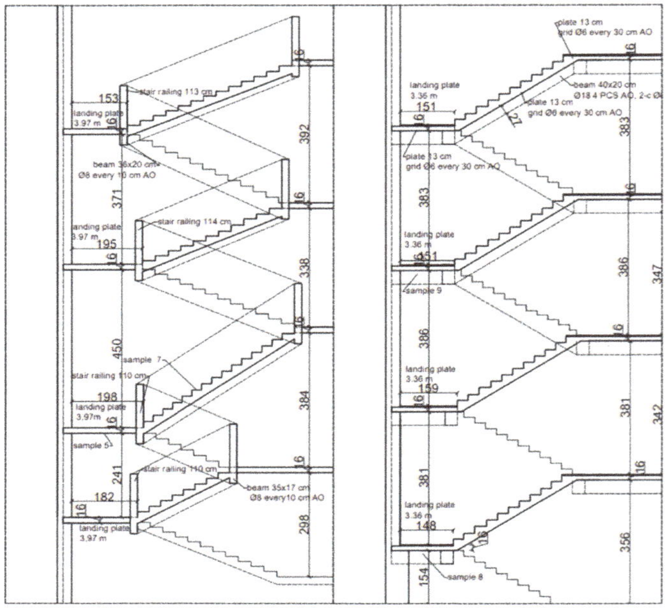

(a) (b)

Figure 2. A view of staircases (**a**) staircase K1, (**b**) staircase K2.

Figure 3. A view of a staircase with a sample 8 location.

Another tested element of vertical communication was the K-2 staircase, which is fully monolithic. It was built by the P-3 slab (13 cm thick, with 16 cm thick landing and the following layers: 13 cm reinforced concrete slab and 3 cm of finishing layers). On both sides, it rests on 2 B-3 reinforced concrete stringers. In the level of the platforms the slabs rest on the new supporting structure, and the other side on the wall. They were connected with a massive B-3.1 reinforced concrete beam in the space between the running plates of the landing, probably in order to stiffen the entire structure. The staircase diagram is shown on Figures 2b and 4. Sclerometric and destructive tests were carried out on the concrete of running and landing slabs as well as reinforcement iron (steel) inserts.

Figure 4. A view of the K2 staircase.

It was not necessary to carry out a statistical analysis in this example and plan the experiment, because the samples were made to verify those presented in previous studies. The general aim was not a new experiment, but to determine the parameters of the existing structure, which was associated to limit the number of taken samples.

On Figure 2b. The diagram of K2 staircase indicates the place of collecting the research samples. Sample No. 8 was taken from the landing slab. A core sample was taken (Figure 5) and then tested in a strength press. Representative sample No. 9 was taken from the landing.

Figure 5. Core concrete sample No. 8 (borehole diameter 98 mm).

The tests of taken samples were carried out in the Pilot 4 testing machine in accordance with the applicable standards. The obtained results are presented in Tables in Section 3.

2.2.2. Ceramic K5 Staircase

In the first decades of the 20th century, ceramic bricks were commonly used, the so-called clinker—with compressive strength Kc = up to 640 kg/cm^2, the so-called class I masonry with Kc = up to 160 kg/cm^2 (16 MPa), class II with Kc = ~100 kg/cm^2 (10 MPa) and sand-lime with Kc = ~180 kg/cm^2 (18 MPa). In fact, the strength class of the bricks was much higher, which is confirmed by most of the research carried out today. The current condition of masonry structures depends on the conditions in which they were used, as well as whether the maintenance and repair procedures have been carried out or not. The collected samples of bricks (walls) from the walls of the building [32] have shown their examination a considerable range of strength, even up to 30 MPa (in a dry condition, i.e., with a mass humidity of $\leq$3%). The average value of this strength was set at 18.1 MPa. A similar result was obtained during laboratory tests [28], f = 28 MPa. The ceramic structures of the staircases did not show any moisture content, so the level of moisture in the bricks corresponded to the dry condition (acceptable humidity).

This part of article describes tests of the ceramic and vaulted structural elements of the K5 staircase, which also served as a starting point for the verification calculations. The load-bearing structure of the staircase consists of ceramic running (bricks and mortar), landing and landing plates based on walls and iron (steel) beams. The main subject of the research and analysis of the structure in question if a brick was a running plate (vault with a thickness of $\frac{1}{2}$ brick—12 cm) with an arc arrow up to 24 cm, shown above (Figures 6 and 7), on which the places of sampling for strength research were marked. Representative sample no. 1 (core borehole) was taken in the course of the staircase.

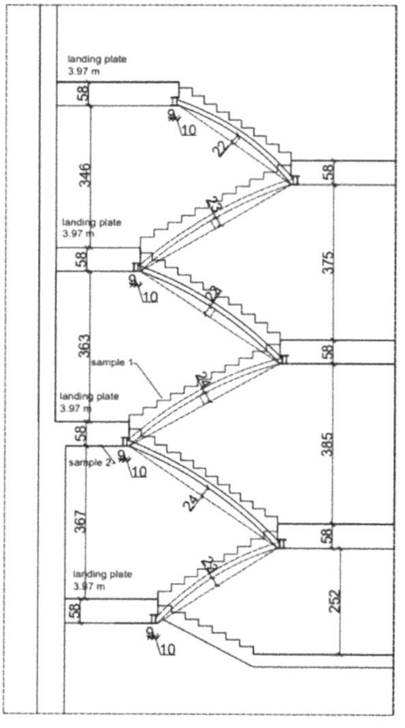

Figure 6. A view of K5 staircase.

Figure 7. A view of a staircase with sample 1 location.

Representative sample No. 2 was taken from the landing (Figure 8). The sample tests were carried out in the Pilot 4 testing machine in accordance with the applicable standards. The obtained results are presented in Table 1.

Figure 8. Bitumen sample (**a**) and a core cross-section (borehole diameter 98 mm) (**b**).

Table 1. The strength test results for samples no. 1 and 2.

Sample No.	Sample Weight [kg]	Average Height of the Prepared Sample [mm]	Average Length of the Prepared Sample [mm]	Compression Strength f [N/mm^2]
1 (average-horizontal test)	1.387	98	98	28.05
1a	1.385	97	98	27.09
2a	1.385	98	97	28.15
2b	1.387	97	99	28.20
1d	1.387	98	98	28.23
1e	1.387	98	98	28.31
1f	1.387	98	98	28.05

Table 1. *Cont.*

Sample No.	Sample Weight [kg]	Average Height of the Prepared Sample [mm]	Average Length of the Prepared Sample [mm]	Compression Strength f [N/mm^2]
2 (average-vertical test)	0.867	122	98	5.00
2a	0.872	123	99	5.05
2b	0.863	121	98	4.93
2c	0.866	122	97	4.83
2d	0.867	122	97	5.09
2e	0.869	123	99	5.02
2f	0.863	121	97	5.09

Slag/the so-called slag mixed with tar, for which the bulk and tapped bulk densities were determined to be 0.89 g/cm^3 and 1.08 g/cm^3, respectively.

3. Results

For each staircase, samples 8a–f were taken in the running plate and samples 9a–f in the landing plate. Results are presented in this section.

3.1. K5 Ceramic Staircase

The obtained results are presented in Table 1—there are two representative samples (samples 1 were taken within the running plate, samples 2 were taken within the landing plate), and each had results a–f. Tests were done on different sub-samples from the same brick, which were taken in multiple places.

Basic statistics presented in the following figures were carried out for the research results (Figure 9).

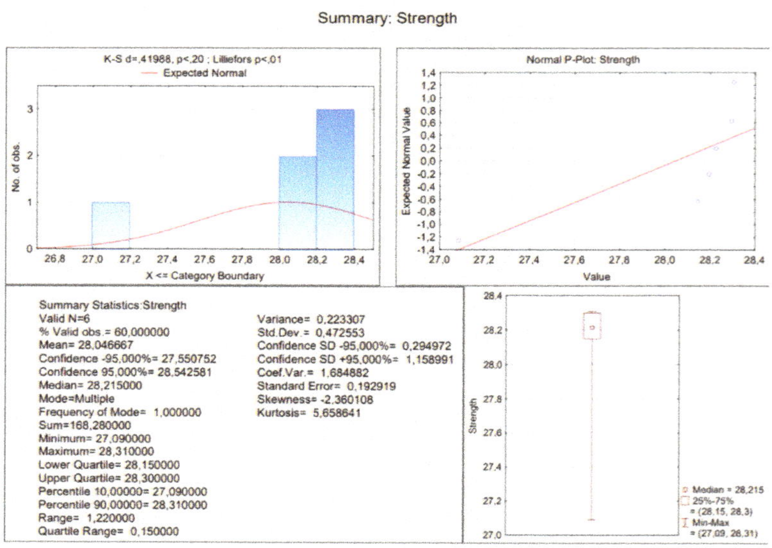

(a)

Figure 9. *Cont.*

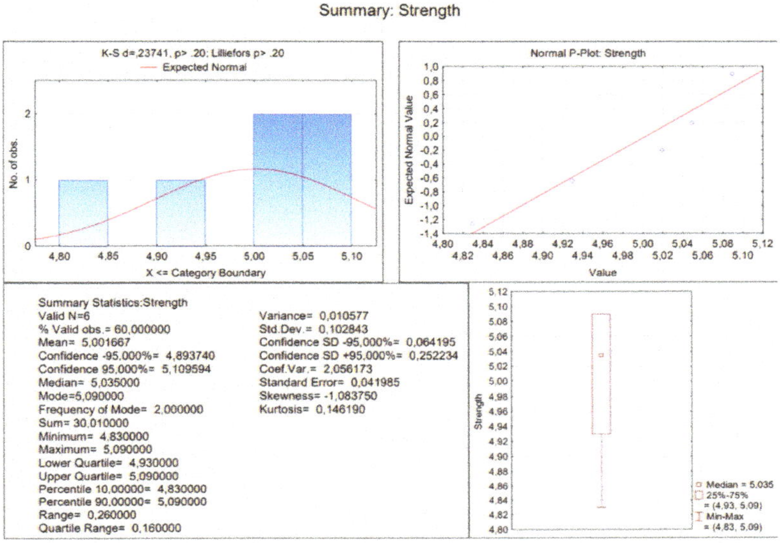

(**b**)

Figure 9. A view of statistics (**a**) samples 1a–1f; (**b**) samples 2a–2f.

The values of internal forces in the vault were determined using Robot Autodesk Structural Analysis. The selected calculation results are shown on Figure 10, which shows an example of the calculation results.

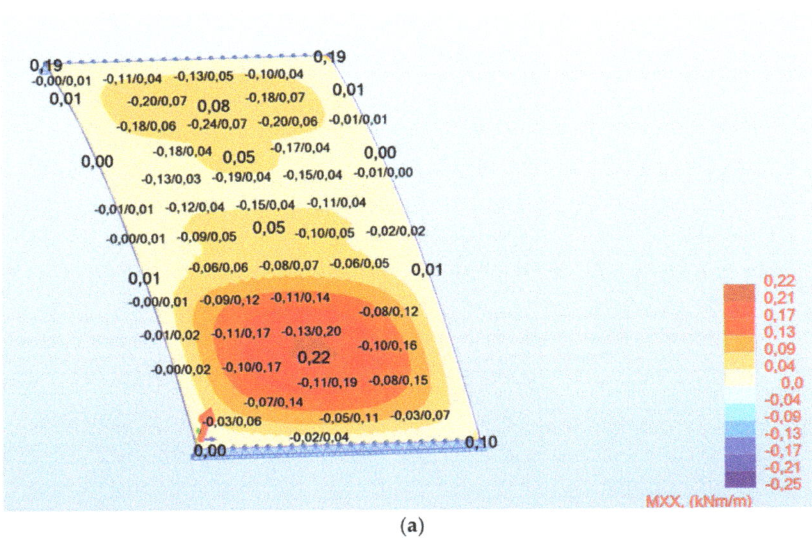

(**a**)

Figure 10. *Cont.*

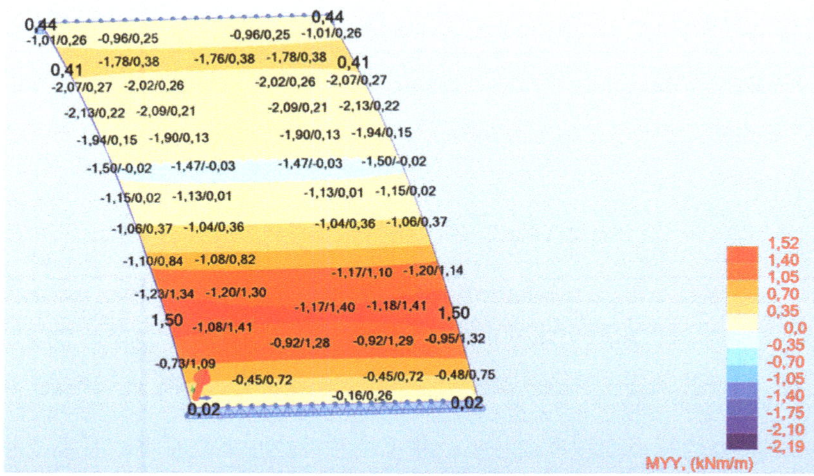

(**b**)

Figure 10. A view of the dialog box (**a**) map of XX moment, (**b**) map of YY moment.

As mentioned in the studies [26,27], the ceramic running plates are vaulted structures and as vaults they did not require separate reinforcement. To calculate the stresses and check the bearing capacity of the above-mentioned, the structure was verified with the aforementioned Robot Autodesk Structural Analysis program. Whereas the calculation results were verified with the Rama R2D2 Intersoft program and the proprietary algorithm [26], created in the Mathcad 7 program, based on standards [20,21,24]. In the study [28], the compressive strength of bricks was determined at the level of 28 MPa, while in the study [32] at an average of 18.1 MPa. Finally, the compressive strength of bricks was assumed for the verification calculations, equal to 15 MPa and mortar of the class (brand) of 5 MPa (cement-lime mortar, as it was present here). The principle of the correct shaping of the vaults was to adjust their shape (the arch height f) in such a way that the pressure line was within the contour of the loaded element. In this case, we deal locally with tensile stresses in the ceramic cross-section of the running plate, causing the pressure line to exceed the cross-section.

3.2. K1 Reinforced Concrete Staircase

The obtained results are presented in Table 2—there are two representative samples (samples 1 were taken from the running plate, samples 2 were taken from the landing plate), and each had results a–f. Tests were done on different sub-samples from the same concrete, which were taken in multiple places.

Table 2. Strength test results for samples no. 5a–f and 7a–f.

Sample No.	Sample Weight [kg]	Average Height of the Prepared Sample [mm]	Average Length of the Prepared Sample [mm]	Strength on Compression f [N/mm^2]
5 (average)	1.480	98	85	36.4
5a	1.45	99	83	36.3
5b	1.5	96	86	35.9
5c	1.49	97	86	37
5d	1.48	98	84	36.5
5e	1.48	98	86	36.4
5f	1.46	99	84	36.2
7 (average)	0.867	98	98	22.3

Table 2. *Cont.*

Sample No.	Sample Weight [kg]	Average Height of the Prepared Sample [mm]	Average Length of the Prepared Sample [mm]	Strength on Compression f [N/mm^2]
7a	0.859	95	96	22.3
7b	0.873	100	94	22
7c	0.862	99	97	22.3
7d	0.865	97	99	23
7e	0.866	98	99	21.9
7f	0.875	98	100	22.5

Basic statistics presented in the following figures (Figure 11) were carried out for the research results.

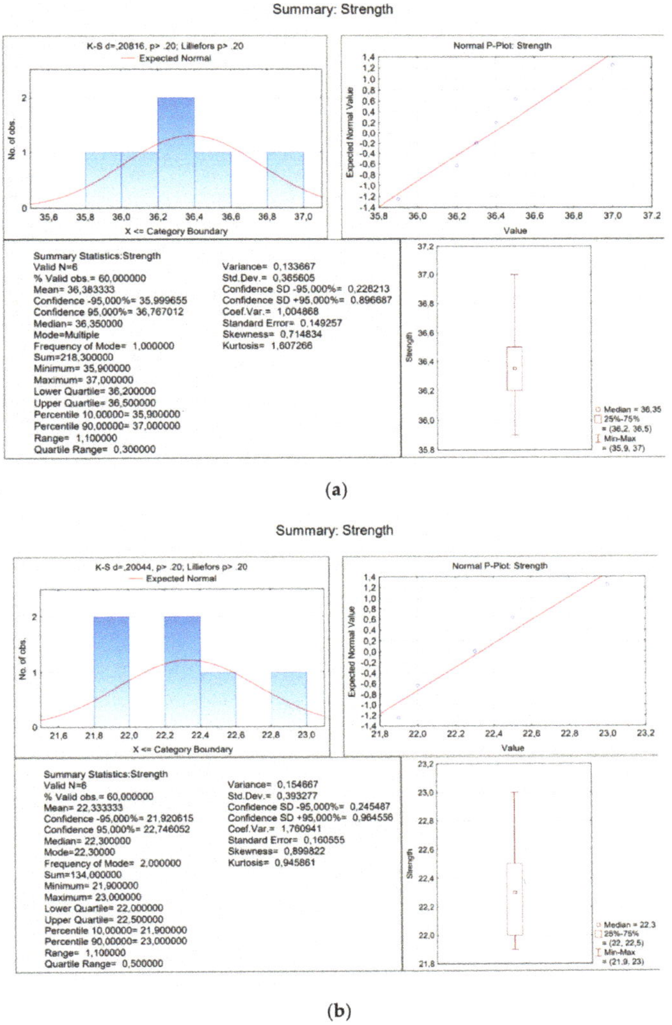

Figure 11. A view of statistics (**a**) samples 5a–f; (**b**) samples 7a–f.

Strength tests of 3 samples (tested in [30]) of concrete used for the structure of stairs in compression according to [30] has shown its very low class in relation to the current requirements B10 ÷ B12.5; therefore, it was initially considered that these structures had insufficient bearing capacity of running plates and landings and should be dismantled according to the owner and project manager. The subsequent research carried out on a much larger number of samples, taken from various places of the staircase, showed higher strength parameters of concrete; therefore, as already mentioned above, the verification calculations were carried out for two concrete classes B15 (C12/15) and B12.5 (without modern concrete). The calculation results obtained for the concrete class B12.5 were assumed as "safer", despite the fact that the tests indicated a much higher class. In [26], the actual number of reinforcement bars was compared to the computationally required amount. This required an additional, more detailed diagnosis on the site before making a decision to leave the existing condition for further use or to supplement the reinforcement.

The structure of the stairs is made of a running plate (13 cm thick) supported on the longitudinal wall of the staircase and a running string, which is also a reinforced balustrade. In the assumptions for the calculations, it was assumed that the balustrade is entirely a load-bearing stair cheek beam. It was also assumed that the P-1 running plate is freely supported on both sides, although in fact it can be fixed on one side in the above-mentioned "cheek". Sclerometric and destructive tests were carried out on the concrete of running and landing slabs, as well as reinforcement iron (steel) inserts. Such an assumption would significantly reduce the span moment, the amount of deflection and the computationally required amount of reinforcement bars. The performed verification calculations showed that the inventoried number of reinforcing bars should be sufficient to transfer the moment, but their spacing does not meet the current code requirements [22], because it is too large. The permissible maximum bar spacing according to PN [22] should be the smaller of the two values: 25 cm or 1.2 h = 15.6 cm, i.e., much smaller than the actual one, set at 24 cm. In this situation, it was proposed to strengthen the boards from the bottom with one layer of PBO mesh applied on the mineral matrix. General requirements for the reinforcement of ceramic and concrete structures with the use of composite materials are given in point 3. The P-2 landing slab is supported on the wall and on the landing beam. The spacing (17 cm) and diameters of the bearing bars (ø8—the equivalent of S235 class steel) are computationally and structurally acceptable here, and only these values should be confirmed directly on the construction site before starting the works. The permissible maximum bar spacing according to PN [22], as stated above, should be the smaller than the two values 25 cm or 1.2 h = 15.6 cm (the actual spacing of ~17 cm is close to the permissible values). If such bar diameters and spacing are confirmed, it will be possible to leave the structure of the plates without additional reinforcements. The B-1 cheek beam, which is also a stair railing (balustrade), has a sufficient load-bearing capacity assuming the work of the entire cross-section, i.e., about 100 cm high.

The problem was the not fully inventoried landing beam B-2, loaded with two large, concentrated forces originating from the reaction of the B-1cheek beams. With the existing cross-section, the bottom reinforcement in the form of 8 bars ø18 (equivalent to steel of A-I class) would be required, while only 2 such bars were inventoried. In the period when the staircases structures has been made, the concentration of reinforcement was used, which would be unacceptable at the present time (sketch below according to [33]—Figure 12, 12ø30 in cross-section with b = 20 cm).

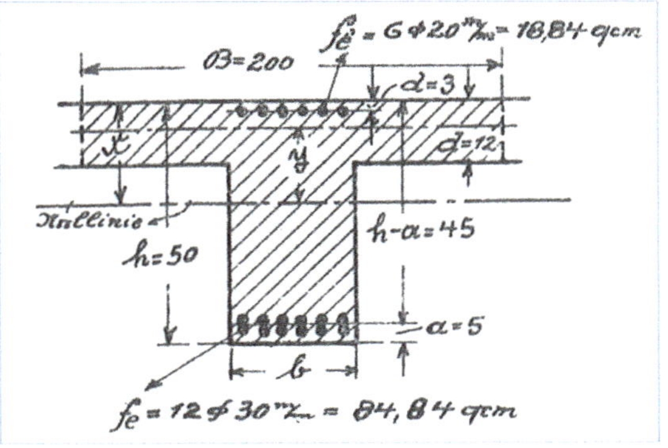

Figure 12. A view of historical structures reinforcement. [33].

At the present times, the width of the beam could not be less than 42 cm. It cannot be ruled out that the existing reinforcement of the B-2 beam has not been fully inventoried or that there is an additional iron (steel) profile inside it. The cross-section of two ø18 iron inserts is F_a = 5.05 cm^2 transfers the moment of ~47.5 kNm with a possible maximum of 177.15 kNm (the required reinforcement is 8ø18 with F_a = 20.36 cm^2). Such a large bending moment could be transferred by the parallelepiped I 200 × 200 (h = 200 mm and b = 200 mm), produced since 1902, which could easily fit into the cross-section of the B-2 beam.

3.3. K2 Reinforced Concrete Staircase

The obtained results are presented in Table 3 and Figure 13—there are two representative samples (samples 1 were taken from the running plate, samples 2 were taken from the landing plate), and each has results a–f. Tests were done on different sub-samples from the same concrete, which were taken in multiple places.

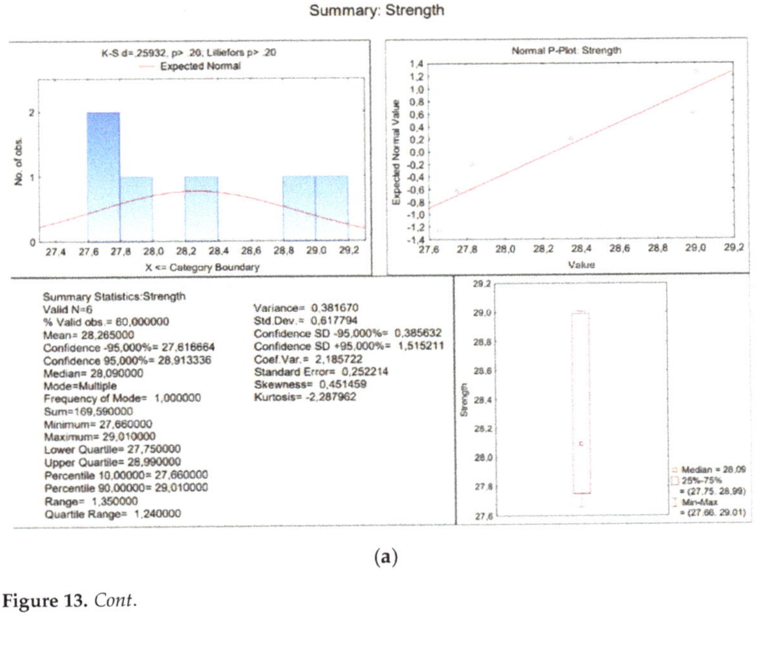

(a)

Figure 13. *Cont.*

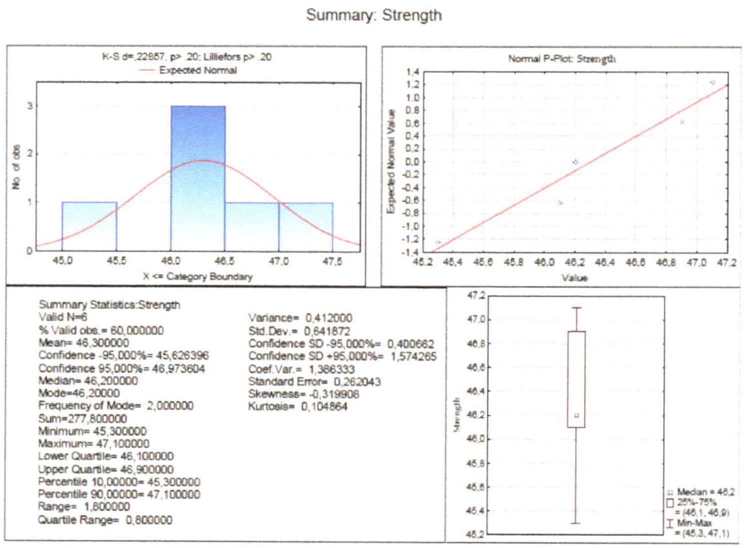

(**b**)

Figure 13. A view of statistics (**a**) samples 8a–f; (**b**) samples 9a–f.

Table 3. Strength test results for sample no. 8, 9.

Sample No.	Sample Weight [kg]	Average Height of the Prepared Sample [mm]	Average Length of the Prepared Sample [mm]	Strength on Compression f [N/mm^2]
8- average	1.45	98	80	28.27
8a	1.46	99	78	28.35
8b	1.44	96	79	29.01
8c	1.47	95	82	27.66
8d	1.42	97	80	28.99
8e	1.46	99	87	27.83
8f	1.45	98	76	27.75
9- average	1.19	98	70	46.3
9a	1.2	98	70	47.1
9b	1.23	96	71	45.3
9c	1.15	99	73	46.2
9d	1.34	97	72	46.9
9e	1.13	98	69	46.1
9f	1.17	99	67	46.2

The P-3 plate rests on both sides of the B-3 running beams' cheeks and it cannot be ruled out that it is fixed in them. For this slab, the reinforcement was inventoried in the form of 7 bars ø8/per 1 m width (equivalent of A-I class steel) with F_a = 3.59 cm^2. This spacing (every 14 cm) is sufficient to carry the expected loads, therefore no additional reinforcements are provided for here. The situation is similar with the construction of the B-3 beam. It was recommended that, prior to the commencement of construction works, the size and spacing of the reinforcement inserts should be confirmed in an amount not less than 4ø16 (equivalent to A-I class steel) with F_a = 7.40 cm^2.

4. A Proposal to Strengthen the Existing Structures

A very interesting element of the research was the assessment of over 100-year-old structures, especially the reinforced concrete. It was a pioneering period in the use of reinforced concrete. Despite the fact that the requirements then for the strength classes of materials, diameters and spacing of reinforcement inserts were significantly different from the present ones, the over 100-year service life of these structures proves that the solutions were properly adopted at the beginning of the 20th century. The authors of the research, wanting to keep the existing structures unchanged (to maintain their authenticity), without additional visible interference, proposed to strengthen them by implementing FRCM (Fibre Reinforced Cementitious Matrix) composite systems using carbon mesh of ceramic structures) and PBO (polyparaphenylene benzobisoxazol) mesh (for reinforced concrete structures). Both types of the nets differed from each other by the terms of strength, density and spacing of the carrying fibers. Carbon meshes were characterized by the same spacing of fibers and the same load capacity in two orthogonal directions (it is important for tensile stresses in two directions in the stair vault—diagrams of the distribution of internal forces and stresses in the vault Figure 10). For PBO meshes, the main direction of the load-bearing capacity was determined and the associated much higher fiber density was associated with it. Both carbon mesh and PBO mesh should be attached to properly prepared substrates in inorganic matrices and, if necessary, anchored with system connectors.

The FRCM technology using carbon mesh and PBO mesh is now more and more commonly used to strengthen masonry and reinforced concrete structures, especially those historic ones. A suitable example of this is central Italy, located in the seismic impact zone, where the strengthening of the structure with the "FRCM technology" allowed the preservation of valuable immovable monuments even after several earthquakes. Such solutions, apart from their strengthening function, should be at the same time not noticeable enough not to disturb the aesthetics of the visual reception of the rooms in which they are used, and thus their functional comfort. The thickness of a single reinforcement layer does not exceed 1 cm, and at the same time it can be an additional fire protection for shallowly laid reinforcement, because without losing its load-bearing parameters, such protections can withstand temperatures of 550 °C to 600 °C. This regards the application of meshes only on mineral matrices, but not to joints using epoxy resins. Additional finishing layers, such as plasters, can be applied on the reinforcement. Before the application of FRCM reinforcements, the pull-off strength research of the substrate should be checked (this applies to the adhesion of future matrices). It should be not less than 1.5 MPa. The test should be carried out using the so-called Tear-off mushrooms by gluing them with epoxy glue to the smooth and cleaned surface, circumferential incision of the substrate and then tearing them off using a tear-off device, e.g., Positech AT-M. After starting the works, the wall load-bearing capacity should be assessed in the places of concentrated loads, i.e., in the places where the stringers, landings and platforms are supported. Such checking and possible strengthening of wall fragments would also occur when new stair structures rest on them. The application of the meshes should be preceded by preparatory works with the use of an appropriate concrete or construction ceramics re-profiling system when the substrate is uneven, peeling and with cavities. The system described above is used to strengthen reinforced concrete elements, mainly in the areas of bending, shear and torsional stresses. Strengthening the stair elements should be performed on the entire surfaces qualified for this purpose with the above-mentioned PBO composite meshes. As is the case with the reinforcement of masonry structures with carbon meshes, the application of PBO meshes must be preceded by consultation with a technical specialist of the system application, who will indicate to the contractor all the detailed requirements for its use in the case in question and confirm in writing that the application of the structure reinforcement system has been completed in accordance with all the requirements set by the manufacturer in the technical data sheets. It should be performed under strict supervision of authorized persons (works manager, supervision inspector).

5. Discussion and Conclusions

A comprehensive diagnostic analysis of a building structure often requires the participation of people with appropriate qualifications and experience, which will allow for appropriate orientation and overall assessment of the problem under consideration [1–14,18,19,35–54]. The construction materials discussed in the article are undoubtedly of valuable historical value, especially as they come from the pioneering period of implementing concrete and iron (steel) connections on the European construction market. Their valuable element is also a different computational and constructional approach to the selection of reinforcement inserts of reinforced concrete elements and their distribution in the concrete cross-section of the elements. Moreover, at the turn of the 19th and 20th centuries, as well as at the beginning of the latter, materials with much lower strength parameters were used, which, in the aspect of current research, may equal the materials currently used, but are not identical to them. A particularly valuable object here are the original ceramic vaulted stairs, for which some concerns may be caused by their load-bearing capacity due to the expected quite high operational load, i.e., 4.0 kN/m^2, but at the turn of the above-mentioned centuries, the dimensioning of the stair structure was taken into account already functional loads of 4.0 kN/m^2 (below is a fragment of calculations from 1910 according to [33]—Figure 14- Nutzlast—400 kg/m^2—including the translation from the German language).

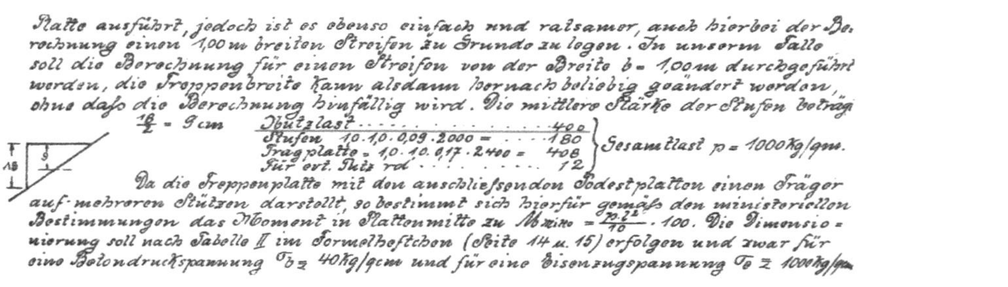
Platte ausführt, jedoch ist es ebenso einfach und ratsamer, auch hierbei der Berechnung einen 1,00 m breiten Streifen zu Grunde zu legen. In unserem Falle soll die Berechnung für einen Streifen von der Breite b = 1,00 m durchgeführt werden, die Treppenbreite kann alsdann hernach beliebig geändert werden, ohne daß die Berechnung hinfällig wird. Die mittlere Stärke der Stufen beträgt $\frac{18}{2}$ = 9 cm

Nutzlast	400	Gesamtlast p = 1000 kg/qm.
Stufen 1,0 · 1,0 · 0,09 · 2000 =	180	
Tragplatte = 1,0 · 1,0 · 0,17 · 2400 =	408	
Für evt. Putz rd	12	

Da die Treppenplatte mit den anschließenden Podestplatten einen Träger auf mehreren Stützen darstellt, so bestimmt sich hierfür gemäß den ministeriellen Bestimmungen das Moment in Plattenmitte zu $M_{max} = \frac{p \cdot l^2}{10} \cdot 100$. Die Dimensionierung soll nach Tabelle II im Formelheftchen (Seite 14 u. 15) erfolgen und zwar für eine Betondruckspannung $\sigma_b \leq 40$ kg/qcm und für eine Eisenzugspannung $\sigma_e \leq 1000$ kg/qcm

Figure 14. The fragment of calculations from publication [33]. Translation below.

"If the calculations are to be carried out for a strip with a width of b = 1.00 m, then the width of the stairs can be freely changed without reducing the correctness of the calculations. The average thickness of the steps 18/2 = 9 cm

operational load—400 kg/m^2

steps—180 kg/m^2

landing plate—408 kg/m^2 total load—p = 1000 kg/m^2

plaster—12 kg/m^2

Since the running plates of the stairs together with the landing plates are a load-bearing structure with many sub-pores, according to the guidelines of the Ministry, the bending moment in the middle of the board should be Mśr = pl^2/10 × 100. Dimensioning should be carried out in accordance with table II in the formula book (page 14 no. 15), specifically for the compressive stress of concrete $\sigma_b \leq 40$ kg/cm^2 and for the tensile stress of iron $\sigma_b \leq 1000$ kg/cm^2."

The position of the authors of the publication is to preserve the historical substance to the highest degree of its authenticity, not mentioning about keeping them in general. It is difficult to carry out reliable research and analyzes in terms of determining the safety level of the structures left behind, the computational approach of which was completely different than the current one. Nevertheless, the inspection of the components of the staircases revealed that, despite over 100 years of exploitation, their deformations or scratches are not visually noticeable. So far, no measurements of this type have been carried out. The performed tests and verification calculations allow for the approval of these structures for further operation after taking into account the assumptions and requirements described above.

Structure materials in historic buildings should not be classified as useless in relation to current regulations and standards, only because of their age. In the considered cases, parallel to the laboratory research of concrete and bricks, non-destructive tests were carried out for them using a "Schmidt hammer" sclerometer, which was calibrated on the basis of samples taken and destroyed in the laboratory. The main purpose of these tests was to check the homogeneity of the material structure on the entire surfaces of these structures, which was also confirmed. The quality of the constructions made in year 1840 and in 1910 should be considered as high, and their technical condition as very good.

Currently, science is trying to look ahead towards the development and research of new materials, technologies, conducting advanced laboratory research, but there are still historical objects that are the national heritage of each region, which should be preserved for future generations in full technical efficiency and the highest degree of authenticity. Universal methods of strengthening, repairs and reinforcement have not yet been developed for them. This article not only attempts to justify such a necessity and to justify the preservation of these objects in a version corresponding to the original, but also to maintain their original function, but in the new conditions of the requirements set by building regulations and standards. What would science be if it did not transfer the results of its research to reality, even the one concerning "obsolete" constructions, one of which is over 180 years old and the other two over 110 years old? These are not standard and typical solutions, because such solutions in historic structures are still rare. We are dealing here with three different, and significantly different, technologies of performance of stairs with currently not used parameters, which have been subjected to multi-format researches and analyzes of materials, indicating simple but innovative reinforcement solutions, full responsibility for their implementation and operational safety. Reinforcing composite materials are already a well-known solution, but it mainly concerns the technology of their fixing with resins. As a way to strengthen the weakened parts of the structure, solutions based on the FRCM technology using carbon meshes and PBO have been proposed, which are described in detail in point 4 of the article, which are still in the research phase, especially in terms of their strength depending on adhesion to substrates and have not yet been fixed in standards.

This approach to the problem of historical buildings is a significant part of the sustainable development strategy, because such buildings can still perform their original functions or can be used for other purposes. In this way, the need to deposit post-demolition waste in landfills that are already disappearing or is eliminated. Such solutions do not require the energy that is needed to produce new materials, transportation and erection of new facilities.

Author Contributions: Investigation, D.B.; Resources, I.T.; Data curation, A.G. All authors have read and agreed to the published version of the manuscript.

Funding: This research received no external funding.

Institutional Review Board Statement: Not applicable.

Informed Consent Statement: Not applicable.

Data Availability Statement: The article is original and previously unpublished.

Conflicts of Interest: The authors declare no conflict of interest.

References

1. ITB Safety Principles Assessment of Reinforced Concrete Structures 1999. Available online: https://www.itb.pl/instrukcje-wytyczne-poradniki (accessed on 6 January 2023).
2. Giergiczny, Z.; Batog, M.; Synowiec, K. Cement and concrete. *Mag. Highways* **2018**, *8*, 62–67.
3. Teomete, E.; Tayfur, G.; Aktaş, E. Estimation of mechanical properties of limestone using regression analyses and ANN. *Cem. Wapno Bet.* **2012**, *17*, 373–389.
4. Maj, M.; Ubysz, A.; Hammadeh, H.; Askifi, F. Non-destructive testing of technical conditions of RC industrial tall chimneys subjected to high temperature. *Materials* **2019**, *12*, 2027. [CrossRef]

5. Kaminski, M.; Maj, M.; Ubysz, A. Chimney cracked reinforced concrete walls as a problem of durability exploitation. In Proceedings of the Fifth International Conference on Structural Engineering (SEMC 2013), Cape Town, South Africa, 2–4 September 2013.
6. Bajno, D.; Bednarz, Ł.; Grzybowska, A. The Role and Place of Traditional Chimney System Solutions in Environmental Progress and in Reducing Energy Consumption. *Energies* **2021**, *14*, 4720. [CrossRef]
7. Malier, Y. *High Performance Concrete*; E&FN Spon: London, UK, 1992.
8. Lai, S.; Serra, M. Concrete strength prediction by means of neural network. *Constr. Build. Mater.* **1997**, *11*, 93–98. [CrossRef]
9. Schifman, J. The Rock Solid History of Concrete 2017. Available online: https://www.popularmechanics.com/technology/infrastructure/a28502/rock-solid-history-of-concrete/ (accessed on 6 January 2023).
10. Dobiszewska, M.; Schindler, A.K.; Pichór, W. Mechanical properties and interfacial transition zone microstructure of concrete with waste basalt powder addition. *Constr. Build. Mater.* **2018**, *177*, 222–229. [CrossRef]
11. Lorenzi, A.; Caetano, L.F.; Chies, J.A.; da Silva Filho, L.C.P. Investigation of the Potential for Evaluation of Concrete Flaws Using Nondestructive Testing Methods. *ISRN Civ. Eng.* **2014**, *2014*, 543090. [CrossRef]
12. Panedpojaman, P.; Tonnayopas, D. Rebound hammer test to estimate compressive strength of heat exposed concrete. *Constr. Build. Mater.* **2018**, *172*, 387–395. [CrossRef]
13. Hoła, J.; Schabowicz, K. State-of-the-art non-destructive methods for diagnostic testing of building structures—Anticipated development trends. *Arch. Civ. Mech. Eng.* **2010**, *10*, 5–18. [CrossRef]
14. Ahmed, A.; Mateo-Garcia, M.; Arewa, A.; Caratella, K. Integrated Performance Optimization of Higher Education Buildings Using Low-Energy Renovation Process and User Engagement. *Energies* **2021**, *14*, 1475. [CrossRef]
15. Ujma, A.; Kysiak, A. Diagnostics of building structure elements with use of thermal imaging camera. *Constr. Optim. Energy Potential* **2015**, 182–190.
16. Wałach, D.; Dybeł, P.; Jaskowska-Lemańska, J. Diagnostics of transport construction structures made of high-performance concrete. *Logistics* **2014**, *6*, 10802–10810.
17. Ahn, E.; Kim, H.; Sim, S.-H.; Shin, S.; Shin, M. Principles and Applications of Ultrasonic-Based Nondestructive Methods for Self-Healing in Cementitious Materials. *Materials* **2017**, *10*, 278. [CrossRef] [PubMed]
18. Hoła, J.; Bień, J.; Sadowski, Ł.; Schabowicz, K. Non-destructive and semi-destructive diagnostics of concrete structures in assessment of their durability. *Bull. Polish Acad. Sci. Tech. Sci.* **2015**, *63*, 87–96. [CrossRef]
19. Hong, S.; Yoon, S.; Kim, J.; Lee, C.; Kim, S.; Lee, Y. Evaluation of Condition of Concrete Structures Using Ultrasonic Pulse Velocity Method. *Appl. Sci.* **2020**, *10*, 706. [CrossRef]
20. *PN-82/B-02001*; Structure Loads. Permanent Loads. PKN: Płock, Poland, 1982.
21. *PN-82/B-2003*; Technological Variable Loads. Basic Technological and Assembly Loads. PKN: Płock, Poland, 1982.
22. *PN-76/B-03264*; Concrete, Reinforced Concrete and Prestressed Structures. Static Calculations and Design. PKN: Płock, Poland, 1976.
23. *PN-B-03264:2002*; Concrete, Reinforced concrete And Prestressed Structures. Static Calculations and Design. PKN: Płock, Poland, 2002.
24. *PN-87/B-03002*; Masonry Structures. Static Calculations and Design. PKN: Płock, Poland, 1987.
25. *Engineering Handbook in Civil Engineering Range*; Part I–III; Professor Stefan Bryła, I.S. (Ed.) Redaction: Warsaw, Poland, 1932.
26. Bajno, D.; Grzybowska, A.; Trzyński, I. Technical Opinion of the Proposed Replacement Historic Staircase Structures in Education and Training Buildings. Gorzów Wielkopolski, Opole, Poland, 2022.
27. Bajno, D. Technical Opinion. Opole, Poland, 2021.
28. Laboratorium, A.-C. Testing of Concrete and Ceramic Elements of Staircases K1, K2 and K5. Chojnice, Poland, 2022.
29. Bajno, D.; Grzybowska, A.; Trzyński, I. Inventory of Staircase Structures Prepared for the Opinion. Chojnice, Poland, 2022.
30. Laboratory Barg. Test Report—Selected Material Tests in the Revitalized Post-Hospital Building at ul. Warszawska in Gorzów (Compression Test of Core Boreholes, Determination of Cover Thickness, Reinforcement Spacing and Diameters). Poznań, Poland, 2021.
31. Laboratory Barg. Research Report—Selected Material Tests in a Revitalized Post-Hospital Building at ul. Warszawska in Gorzów (Staircases: Transitional, K1 and K2). Gorzów Wielkopolski, Poland, 2021.
32. Laboratory Barg. Research Report—Selected Material Tests in a Revitalized Post-Hospital Building at ul. Warszawska in Gorzów (Brick Compression Test, Determination of Cover Thickness, Reinforcement Spacing and Diameters). Gorzów Wielkopolski, Poland, 2021.
33. Toensmann, A. *Der Eisenbetonbau von A. Toensmann Zivil-Ingenieur; Eisenbeton*; Akademie Wismar a. Ostsee: Berlin, Germany, 1910.
34. Czapliński, K. *Old Iron Alloy Products*; DWE: Opole, Poland, 2009.
35. Alsaleem, F.M.; Abiprojo, R.; Arensmeier, J.; Hemmelgarn, G. HVAC system cloud based diagnostics model. In Proceedings of the International Refrigerant and Air Conditioning Conference, No 1508, West Lafayette, IN, USA, 14–17 July 2014.
36. Bashi, A.; Jilkov, V.; Li, X. Fault detection for systems with multiple unknown modes and similar units and its application to HVAC. *IEEE Trans. Control. Syst. Technol.* **2011**, *19*, 957–968. [CrossRef]
37. Bonvini, M.; Wetter, M.; Sohn, M. An FMI-based toolchain for the adoption of model-base FDD. In Proceedings of the ASHRAE/IBPSA-USA Building Simulation Conference 2014, Atlanta, GA, USA, 10–12 September 2014.
38. Brambley, M.R.; Fernandez, N.; Wang, W.; Cort, K.; Cho, H.; Ngo, H.; Goddard, J. *Self-Correcting Controls for VAV System Faults, Filter/Fan/Coil and VAV Box. PNNL- 20452*; U.S Department of Energy: Washington, DC, USA, 2011.
39. Costa, A.; Keane, M.; Torrens, J.; Corry, E. Building operation and energy performance: Monitoring, analysis and optimisation toolkit. *Appl. Energy* **2013**, *101*, 310–316. [CrossRef]

40. Fan, B.; Du, Z.; Jin, X.; Yang, X.; Guo, Y. A hybrid FDD strategy for local system of AHU based on artificial neural network and wavelet analysis. *Build. Environ.* **2010**, *45*, 2698–2708. [CrossRef]
41. Han, H.; Gu, B.; Hong, Y.; Kang, J. Automated FDD of multiple-simultaneous faults (MSF) and the application to building chillers. *Energy Build.* **2011**, *43*, 2524–2532. [CrossRef]
42. Han, H.; Gu, B.; Wang, T.; Li, Z. Important sensors for chiller fault detection and diagnosis (FDD) from the perspective of feature selection and machine learning. *Int. J. Refrig.* **2012**, *34*, 586–599. [CrossRef]
43. Hao, X.; Zhang, G.; Chen, Y. Fault-tolerant control and data recovery in HVAC monitoring system. *Energy Build.* **2005**, *37*, 175–180. [CrossRef]
44. Katipamula, S.; Brambley, M.A. Review article: Methods for fault detection, diagnostics, and prognostics for building systems—A review, Part I. *HVAC&R Res.* **2005**, *11*, 3–25.
45. Katipamula, S.; Brambley, M.A. Review article: Methods for fault detection, diagnostics, and prognostics for building systems—A review, part II. *HVAC&R Res.* **2005**, *11*, 169–187.
46. Lauro, F.; Moretti, F.; Capozzoli, A.; Khan, I.; Pizzuti, S.; Macas, M.; Panzieri, S. Building fan coil electric consumption analysis with fuzzy approaches for fault detection and diagnosis. *Energy Procedia* **2014**, *62*, 411–420. [CrossRef]
47. Narayanaswamy, B.; Balaji, B.; Gupta, R.; Agarwal, Y. Data driven investigation of faults in HVAC systems with model, cluster and compare (MCC). In Proceedings of the 1st ACM Conference on Embedded Systems for Energy-Efficient Buildings, New York, NY, USA, 3–6 November 2014; pp. 50–59.
48. Nassif, N.; Moujaes, S.; Zaheeruddin, M. Self-tuning dynamic models of HVAC system components. *Energy Build.* **2008**, *40*, 1709–1720. [CrossRef]
49. O'Neill, Z.; Pang, X.; Shashanka, M.; Haves, P.; Bailey, T. Model-based real-time whole building energy performance monitoring and diagnostics. *J. Build. Perform. Simul.* **2014**, *7*, 83–99. [CrossRef]
50. Song, Y.H.; Akashi, Y.; Yee, J. A development of easy-to-use tool for fault detection and diagnosis in building air-conditioning systems. *Energy Build.* **2008**, *40*, 71–82. [CrossRef]
51. Srivastav, A.; Tewari, A.; Dong, B. Baseline building energy modeling and localized uncertainty quantification using Gaussian mixture models. *Energy Build.* **2013**, *65*, 438–447. [CrossRef]
52. Wang, S.; Zhou, Q.; Xiao, F. A system-level fault detection and diagnosis strategy for HVAC systems involving sensor faults. *Energy Build.* **2010**, *42*, 477–490. [CrossRef]
53. Zhu, Y.; Jin, X.; Du, Z. Fault diagnosis for sensors in air handling unit based on neural network pre-processed by wavelet and fractal. *Energy Build.* **2012**, *44*, 7–16. [CrossRef]
54. Zogg, D.; Shafai, E.; Geering, H. Fault diagnosis for heat pumps with parameter identification and clustering. *Control. Eng. Practice* **2006**, *14*, 1435–1444. [CrossRef]

materials

MDPI

Article

New Ceramic Tiles Produced Using Old Technology Applied on Historic Roofs—Possibilities and Challenges

Krzysztof Ałykow [1], Łukasz Bednarz [2,*], Magdalena Piechówka-Mielnik [2], Magdalena Napiórkowska-Ałykow [1] and Michał Krupa [3]

[1] Team of Civil Engineers, 59-800 Luban, Poland
[2] Faculty of Civil Engineering, Department of Building Structures, Wrocław University of Science and Technology, 50-370 Wroclaw, Poland
[3] Faculty of Architecture, Chair of Housing Environment, Cracow University Technology, 31-155 Cracow, Poland
* Correspondence: lukasz.bednarz@pwr.edu.pl

Abstract: In the case of historic buildings, especially those under protection, it is important to replace elements of the roof covering, while maintaining current technical standards, to meet the requirements of the conservator. The authors of the article present alternatives to commonly used solutions, based on their experience with replacing historic building roofing with ceramic tiles made according to the production and firing technology of the nineteenth century. They emphasize that the correct/specialized restoration of existing tiles in a building makes it possible to preserve and reuse them, which is in line with the principles of historic preservation. However, due to the preservation of the roof tiles, it is not always possible to revitalize them. As a solution to the problem, the use of clay roof tiles manufactured according to 19th-century firing technology, including handmade methods, is presented, which preserves the geometry of the historic roof tiles. The approach presented by the authors meets both the requirements of conservation theory and the building standards for roofing elements. Although it is much more expensive than the solutions currently commonly used that result from modern technical requirements and tile-manufacturing technology, in the case of objects of significant cultural heritage, it is a solution that meets modern technical requirements while not compromising the original appearance of the monument.

Keywords: ceramic tiles; historic roofs; old technology; conservation; heritage

Citation: Ałykow, K.; Bednarz, Ł.; Piechówka-Mielnik, M.; Napiórkowska-Ałykow, M.; Krupa, M. New Ceramic Tiles Produced Using Old Technology Applied on Historic Roofs—Possibilities and Challenges. *Materials* **2022**, *15*, 7835. https://doi.org/10.3390/ma15217835

Academic Editor: Pedro J. Sánchez-Soto

Received: 27 September 2022
Accepted: 4 November 2022
Published: 6 November 2022

1. Introduction

The frequent replacement of old elements with new elements in historic buildings (such as ceilings, roof trusses, window and door joinery, plaster, roof coverings, insulation, the body of the walls, etc.) irretrievably deprives them of their historical and scientific value, causing the problem that after a thorough "revitalization", what remains is a "candy" new building that merely imitates the original one. This is not only contrary to the principles of protection that constitute the rule of law [1–3], but it deprives us of a material legacy that is a visible sign of the activity and presence of the generations before us.

In particular, when it comes to roofs, it is common to approach original tiles as merely a technological element that protects the roof of the building from weather, which is usually understandable, but in the case of historic buildings covered with historic handmade tiles, it is most often not correct [4–6]. Therefore, if the value of a monument is determined by its originality, why are the original roofing materials removed in most cases and replaced with modern machine-made ones?

This paper presents the authors' proposal on how to simultaneously meet the technical criteria arising from contemporary technical standards while at the same time satisfying the requirements arising from the theory of historic preservation. This is of great importance, especially in the case of buildings of significant historical and artistic value and those under the protection of the Office for the Protection of Monuments.

From the point of view of the ceramic tile roofing technology commonly available on the market and widely used today, it would seem impossible to produce a roof covering manufactured in accordance with modern building standards while at the same time meeting the requirements of monument conservation theory. Therefore, ensuring that the proper requirements of modern building standards are met while ensuring authenticity and preserving the traces of historic alterations and transformations is an often-overlooked aspect of the activities of today's architects and engineers.

The basic classification of roof tiles is based on the material from which they are made, that is, ceramic tiles (fired from clay) and cement tiles (made from a combination of cement and sand with additives). Ceramic tiles are slightly lighter than cement tiles and are available in a wider range of colors. However, the natural red brick shade is the most popular because it blends best with a variety of facades and surrounding areas.

Of today's clay roof tiles, the most common shapes used are the Dutch (also known as S-shaped), overlapping, Marseille, or plain tile (Figure 1). Flat roof tiles are available as clay and cement tiles. A modern solution is the photovoltaic tile, which makes it possible to produce electricity. This type of tile is a novelty on the construction market, but is not recognized by the conservation community.

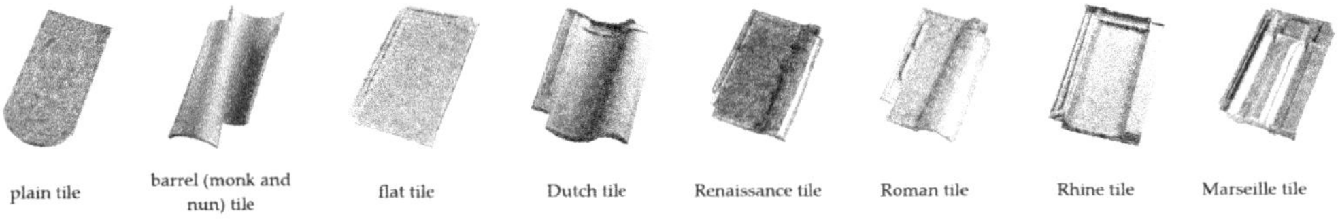

Figure 1. Examples of different roof tiles.

Over the years, as a result of UV radiation, rain, and other atmospheric conditions, tiles can become stained, lose color, and thus start to look unattractive (Figure 2a,b). To renovate the tiles, it is not necessary to replace the entire roof immediately. It is also possible to renovate them, waterproof the surface, and strengthen the structure with suitable chemicals based on silane and silicate particles.

The most effective method is to clean the tiles using a pressure washer (Figure 3a) with a rotary nozzle at a pressure of min. 220 bar. Using this method, loose dirt, moss, and old, poorly adhering coatings can be removed. Particular care should be taken to thoroughly remove moss from the tiles.

The most common means of restoring historic roof tiles after cleaning are silicate-based agents (Figure 3b), which do not add a new color to the ceramics (they are colorless), thus achieving colorless protection while at the same time impregnating and closing the pores in the old tiles. This action also protects them from moisture. This is particularly important under the influence of changing weather conditions. By closing the pores in the ceramics, the silicates create a smoother finish and thus inhibit moss growth on the roof. Tile waterproofing treatments are available in gloss, satin, and matt finishes. Unfortunately, not all roof tiles can be revitalized, especially if their structure and mechanical properties disqualify them.

(**a**)

(**b**)

Figure 2. Examples of historic roof coverings after many years of use: (**a**) handmade late medieval and Renaissance roofing tile; (**b**) machine-made tiles approximately 100 years old.

(**a**)

(**b**)

Figure 3. Revitalization of historic roof tiles: (**a**) cleaning; (**b**) maintenance.

As the problem of preserving old tiles is very complex and costly, the most common route is to choose a new material to mimic the old one. The use of new machine-made roofing tiles has many negative consequences:

- Changes to the appearance of the roof covering;
- The obliteration of the original arrangement of tiles, often of different shapes, built into the roof slope and constituting evidence of its transformation relevant from the point of view of monument documentation;
- The necessity of leveling the roof slope for new tiling, resulting in the incorporation or, worse, the removal of original carpentry elements or their fragments;
- The need to apply new layers, including vapor-permeable foil, which sometimes leads to a disruption of the microclimate within the loft and acceleration of the biological corrosion process of the original elements of the roof trusses;
- A reduction in the weight of the roof covering, which in the case of high roofs (e.g., churches) and in places of contact with high partitions (e.g., church towers) often results in the tiles being torn out of the roof slope despite their proper fixing.

For roof coverings of historic buildings, it is important to produce adequate documentation that includes measurements of the geometry of the existing tiles and, if necessary, their layout on the slope of the roof. An example of such measurements of the geometry of historic roof tiles to help produce replacements is shown in Figure 4.

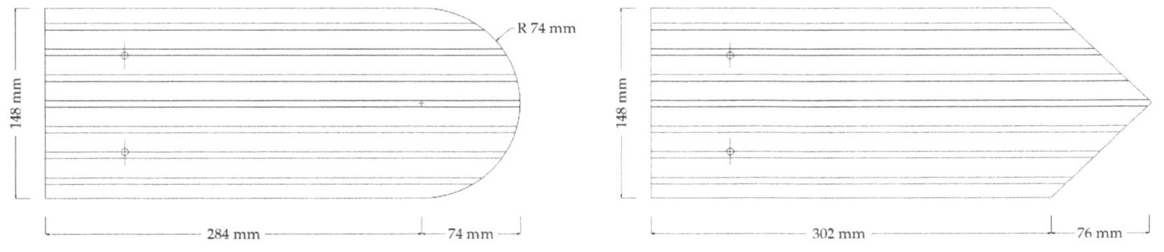

Figure 4. Geometry of historic roof tiles.

It is equally important to determine the degree of technical degradation of historic roofs and, in the case of undamaged roof tiles, how they should be cleaned, maintained, and possibly reinforced. The next step is to indicate the extent and manner of replacing the roofing elements.

Tiles that are part of the restoration and replacements of original ones should be adapted in a way that does not obliterate the original architectural layout and made in such a way as to meet the technical requirements, while allowing specialists to recognize the original and secondary elements, according to the requirements of monument conservation.

According to the authors, one of the most interesting solutions is the use of ceramic tiles made according to 19th-century firing technology, hand-formed and fired in traditional coal-fired Hoffman furnaces.

Based on the results of the laboratory tests carried out by the authors and on the authors' experience with repairs/maintenance of the roof coverings of historical buildings, the article shows that solutions are possible and available that meet both technical and conservation criteria at the same time. This is very important in order to avoid falsifying the historical appearance of a building, which is unacceptable from the point of view of heritage protection.

2. Materials and Methods

To compare the physical and mechanical performance of old and new tiles that look the same, we tested the methods for making historic clay roof tiles, dated as min. 100 years old. The historic tiles were compared with new tiles made with traditional technology, but with up-to-date technical parameters. The tests included determining water absorption and permeability, assessing frost resistance, and a bending strength test. There were 52 tiles used in the tests, of which one was damaged during transport—sample No. 051 cracked.

Before testing, the material was visually assessed. It was found that:

- The tiles came in various sizes and shapes;
- The age of the tiles was estimated to be around 100 years;
- Traces of various mortars were observed on many tiles;
- The tiles were heavily and moderately soiled;
- Numerous roof tiles were chipped;
- There were numerous traces of biological corrosion on the right (upper) surfaces of the tiles, which was contributed to by moss and algae growth (visible, for example, on sample Nos. 010, 029, and 038);
- One of the tiles (sample No. 051) was cracked.

2.1. Water Absorption Test

A total of 30 samples of the dismantled clay tiles (No. 021 to No. 046, No. 048 to No. 050, and No. 052) were selected from the tiles supplied for the test, dried for 24 h in an oven at 110 °C +/− 5 °C (Figure 5a), and weighed to the nearest 1 g. Then, they were subjected to a water absorption test (Figure 5b) according to the procedure in Appendix A of the current standard in [7]. The results of the n_m mass absorption were between

$n_{m\ min}$ = 9.7% (sample No. 031) and $n_{m\ max}$ = 15.4% (sample No. 029), as shown in Figure 6. The average absorption for all 30 tiles was 13.9%.

(a) (b)

Figure 5. Absorption test: (**a**) drying of tiles to constant weight at a temperature of 110 °C +/− 5 °C; (**b**) roof tiles in a chamber with 100% humidity.

Unfortunately, there are no standards for the maximum water absorption values of clay plain tiles. According to the world's operating major tile manufacturers, the maximum water absorption of approved clay roof tiles should not exceed 10%. This water absorption condition was not met by 28 of the 30 samples.

2.2. Frost Resistance Test

The frost resistance test was performed according to the current standard in [8]. Before the frost resistance test, 6 tiles (Nos. 021, 024, 029, 031, 038, and 050) were selected from the 30 tiles (No. 021 to No. 046, No. 048 to No. 050, and No. 052; tile No. 047 was a broken tile) on which the tile water absorption test was performed. Tiles that were free from unacceptable damage in the test and characterized by their minimum, maximum, and average water absorption (2 pieces each) were selected and subjected to frost resistance tests according to the procedure presented in the standard in [8] (Figure 7). Table 1 summarizes and compares the tiles before and after the frost resistance test.

The following were observed on tiles subjected to the frost resistance test after the test: surface scratches, peeling, delamination, and spalling, examples of which are shown in Figure 8a–c.

According to the standard in [9], the clay roof tiles used in Central Europe should be Class 1; that is, they should not show any of the types of damage specified in Table 1 of the standard in [8] after 150 freeze/thaw cycles. The condition of resistance to frost was not fulfilled.

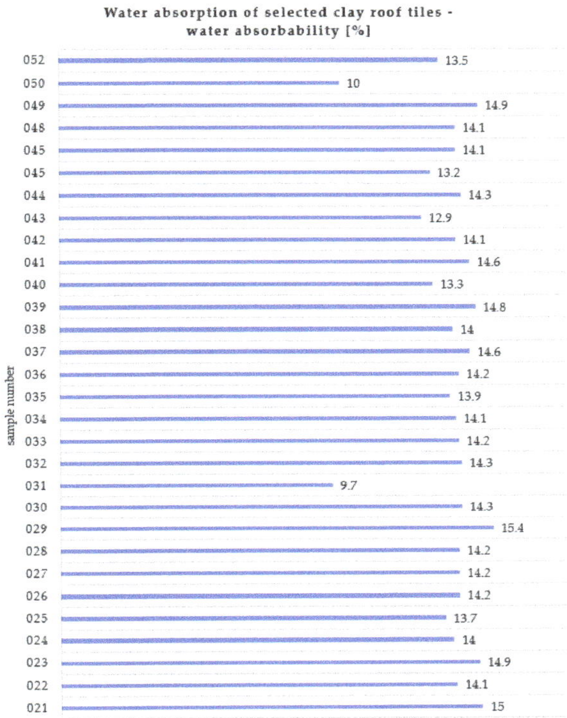

Figure 6. Water absorption of selected historical clay roof tiles.

Figure 7. Roof tiles in a climate chamber.

Table 1. Comparison of the visual condition of the selected tiles before and after the frost resistance test.

Tile No.	Appearance Before Frost Resistance Testing	Appearance After Frost Resistance Testing
021		
024		
029		
031		
038		
050		

(**a**) (**b**) (**c**)

Figure 8. Frost resistance test: (**a**) surface damage observed after the test; (**b**) edge damage observed after the test; and (**c**) damage observed after the test.

2.3. Permeability Test

The permeability test was performed according to the current standard in [10]. After selecting 10 samples (No. 011 to No. 020) of tiles, the tests were carried out according to the procedure described in the standard in [10] (Figure 9). The tile permeability test consisted of determining the time that elapsed until the first drop of water fell under the pressure of the water column exerted on the upper surface of the tile, under normal atmospheric conditions. The test consisted of lying the tile samples in tap water at room temperature for 48 h ± 4 h. The samples were then dried at 110 °C ± 5 °C to a constant weight. The final step was to cool for 4 h at room temperature. The test lasted no longer than 20 h. The permeation coefficient (IC) was calculated using the relevant formulae.

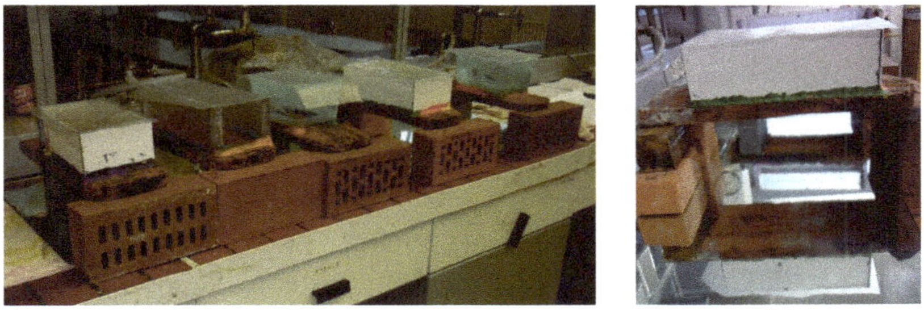

(**a**) (**b**)

Figure 9. Roof tiles during the impermeability test: (**a**) general view; (**b**) view of the underside of tile.

The test results for the individual tiles and the time to the first drop of water (h) are shown in Figure 10. The average time to the first drop of water for the entire test series was $X_{i\ av}$ = 4.98 h. The highest value of the single water absorption coefficient was $ICX_{i\ av}$ = 0.751. The mean permeation coefficient was $ICX_{i\ av}$ = 0.751. The value of the largest single-sample permeation coefficient was ICX_i = 0.988.

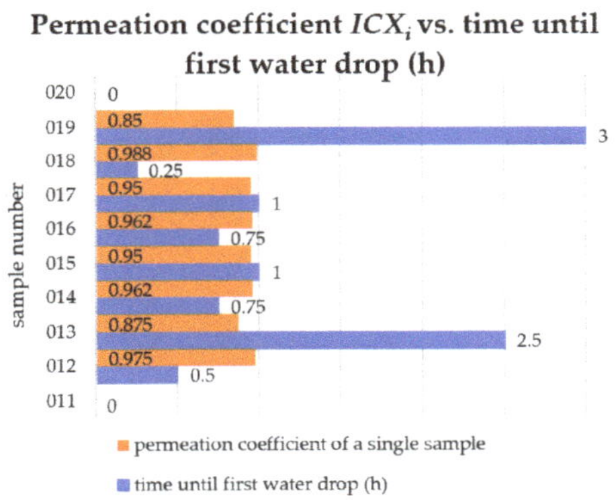

Figure 10. Permeation coefficient.

According to the standard in [9], for standard clay tiles approved for Category 1, the average permeability coefficient of $ICX_{i\ av}$ should be 0.8 or less, and all values of the permeability coefficient of ICXi for individual samples should be 0.85 or less. The condition for the average permeation coefficient $ICX_{i\ av}$ was met. The condition for the values of the permeation coefficients for individual ICX_i samples was not met by seven out of ten samples.

2.4. Flexural Load Capacity Test

The bending resistance test was carried out according to the current standard in [7]. After selecting 10 clay tile specimens (No. 001 to No. 010), the tests were carried out according to the procedure described in the standard in [7] (Figure 11a,b). Figure 12 summarizes the load-bearing results obtained on the selected clay roof tiles tested. The average failure load of F_{av} was 0.24 kN.

(a) (b)

Figure 11. Flexural load capacity test: (**a**) tile in the testing machine during flexural capacity tests; (**b**) tiles (Nos. 001–010) after flexural strength tests.

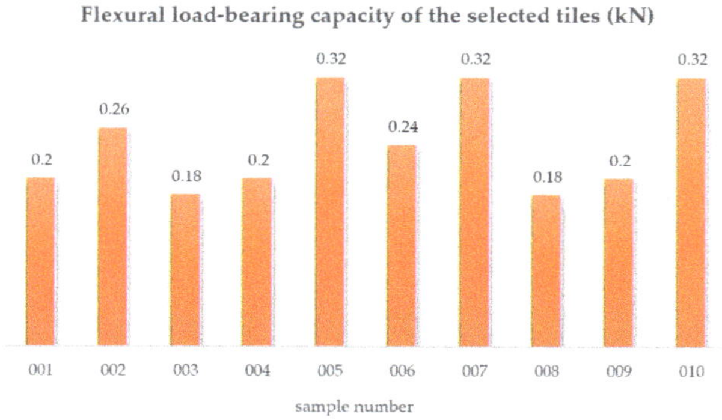

Figure 12. Flexural load-bearing capacity of the selected tiles.

According to [9], for approved plain clay tiles, the bending strength should take a value of min. 600 N (0.60 kN). The bending strength condition was not met in all 10 samples.

In comparison, hand-formed tiles and fired according to the technology of the nineteenth century in coal-fired Hoffman furnaces have the properties shown in Table 2 and meet the requirements of the current standards [7–10].

Table 2. Physical and mechanical properties of the new ceramic tiles produced with the old technology.

Properties	Value
Water absorption	<10%
Permeability	$ICx_i \leq 0.85$
Frost resistance	Frost resistant after 150 cycles
Flexural load capacity	≥1.2 kN

3. Results of Case Study

The laboratory tests carried out for historic clay tiles showed that the roofing material did not meet the modern requirements for reuse in the renovated building. Basic physical–mechanical tests reflected the condition of the historic tiles as a material unsuitable for effective roofing, such as:

- The absorbability (water absorption) condition was not met by 28 out of 30 samples;

- Although the condition regarding the average water absorption coefficient ICX_{iav} was met, the value of the water absorption coefficient for single samples, ICX_i, was not met by 7 out of 10 samples;
- The condition concerning frost resistance was not fulfilled;
- The condition concerning flexural capacity was not fulfilled.

In comparison, new tiles formed by hand and fired according to 19th-century technology in coal-fired Hoffman furnaces, although almost identical to the historic tiles, complied with the current standards (Table 2).

To illustrate the applicability of new ceramic tiles produced with old technology, a short case study of the application of this type of tile in a historic building is presented below. The reference object is the Salt House (Figure 13) of 1539, located in southern Poland [11]. It is located in a space between two rows of medieval defense walls erected before 1220. In 1566, there was a fire in the town, but the building was not damaged besides a few wooden elements. Extended in 1698, the building was used as a storage house for salt and grain until the end of the 18th century. In the 19th century, it was partly converted for use as a prison and the headquarters of the town's fire brigade. The building served as the firemen's headquarters up to the 1990s of the twentieth century.

Figure 13. The Salt House elevation view.

The Salt House is situated on a hill within the chamber of the ramparts, i.e., between the higher and lower lines of the ramparts. This location was due to fire safety considerations, since the high walls effectively separated the building from the flammable buildings of the city, mostly wooden, and also protected it from flooding in the event of rainfall and floods.

The Salt House was built on a rectangular plan measuring 33.8 × 18.4 m from basalt stone and field pebbles bonded with lime mortar with a clay mixture, with the north wall being coextensive with the inner fortification wall. Its numerous but small window openings on the southwest, south, and southeast sides of the town could be used as rifle ranges if necessary. It eventually reached a height of 22 m, accommodating four stories and an attic.

Inside, all of the ceilings received a wooden structure and were supported by wooden columns. The different levels were connected by stairs in the form of ladders supported by ceiling beams. Inside the warehouse, a crane was installed to allow for the loading or unloading of goods. Originally, there must have also been a writer's room and a weighing scale on the ground floor level. The main stone-vaulted gate was located on the northeast side of the building [12–14].

The project documentation created for the renovation envisaged the removal of the existing 19th-century machine-made tiles of no historical or technical value and their replacement with new clay roof tiles, with the application of contemporary layers used in the laying of the new roof, including vapor-permeable foil, which is a commonly used solution [15], but not always correct.

Once the work had started, it was found that on the roof, in addition to tiles dating back to the nineteenth century produced by machine in an amount of approximately 25%, the majority of the remaining were handmade tiles, including some from the period of the building's construction, with four different modes of shaping (flat, segmental, semicircular, and angular) (Figure 14a).

(a) (b)

Figure 14. Different types of roof tiles: (**a**) historical; (**b**) new roof tiles in historic dimensions.

It was also found that on the south slope, the tiles were laid in a scalloped pattern (on the north slope in a lace pattern) with lime mortar fixings with fur and the joints sealed with wooden shackles.

Following the intervention of the building inspector after the renovation work had already begun, the voivodeship conservator of monuments changed his earlier decision to remove the tiles in their entirety and replace them with new ones. He ordered the careful removal, under the additional supervision of a conservator of works of art, of about half of the tiles on the southern slope, ordering that the tiles fixed with original mortar be left on the roof. After analyzing the problem, it was decided to undertake the conservation of the original roofing elements in situ using a hoist and climbing methods (on ropes), while reinforcing the slope by introducing additional battens supporting the original ones. This allowed the original patches and mortar to be preserved and prevented possible damage to the original tiles during their removal and reinstallation after conservation. Only selected roof tiles in poor condition were replaced (Figure 14b). These measures were taken after taking into account all the conservation principles of this type of historic structure [16], while monitoring the static condition of the structure. Some of the procedures proposed in [17] were implemented.

The original decision of the preservation office was also modified, ordering that the original existing roof tiles be cleaned, preserved, and incorporated using traditional mortar fixing. Not all tiles were suitable to undergo this process due to their technical condition.

Instead of the heavily damaged original and 19th-century machine-made tiles, custom handmade tiles resembling the original in appearance, but with much better physical and mechanical properties and with a detailed arrangement of tiles of different shapes on the roof slope reflecting its historic character, were installed (Figure 15).

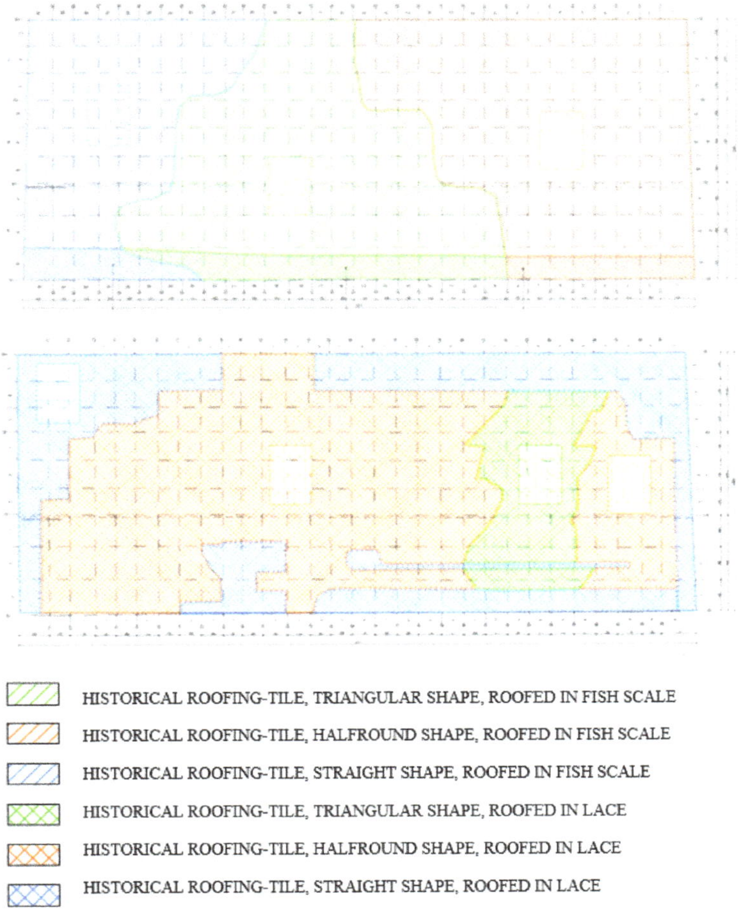

Figure 15. Arrangement of tiles on roof slopes.

4. Discussion

Investigations of historic clay roof tiles have shown significant losses in the quality of the material as a roof covering used in traditional construction. Their failure to meet the requirements for water absorption, permeability, and frost resistance results in poor durability of the material, compared to tiles produced today. Their strengthening through conservation measures, i.e., cleaning and then structural strengthening with chemical preparations based on silicates, is possible, as in the case of the case study presented above, but in the opinion of the authors, due to both the significantly higher costs of this type of work in comparison to ordinary roofing and the greater technological complexity requiring the participation of conservators of works of art in the field of ceramics, it is justified only in the case of historic buildings of particular artistic, scientific, or historical significance. For historic buildings that are not of such significance, but are important from the point of view of individual cultural heritage protection or the protection of a larger urban or rural area [18,19], for example, for the preservation of Genius Loci, the solution proposed by the authors meets the criteria of cultural heritage protection, whose importance is emphasized in the document of the International Council of Monuments and Sites [20].

The use of contemporary handmade roofing tiles, fired in traditional coal-fired Hoffman furnaces, which give the shape, size, and color of historic tiles, makes it possible to achieve a quality roof covering. The importance of this problem can be learned from numerous publications [21–24].

In the case of the presented case study, the change in the method of investment, i.e., the abandonment of the originally designed contemporary tiles in favor of handmade ceramic tiles made according to 19th-century technology in coal-fired Hoffmann cookers and the in situ conservation of part of the original tiles, resulted in an extension of the investment from one to three years and a significant threefold-increase in costs. However, because of a different method of revitalizing the roof covering of the building compared to that originally envisaged, it was possible to preserve the historic elements of the roof. This was achieved in compliance with both contemporary technical requirements and the requirements of the conservation doctrine, thus precisely complying with the requirements recommended by UNESCO [25] for the protection of the historic urban landscape.

This is particularly evident where, along with the original, cleaned, and structurally reinforced roofing elements, new ones have been built in, but produced according to a model and technology corresponding to the historic technology. On the one hand, this approach meets the technical requirements of contemporary standards, while on the other hand, it does not obliterate the original historic appearance of the monument by introducing modern elements that do not harmonize with the original historic substance. An additional and perhaps the greatest benefit of the approach proposed by the authors is that, despite the coherent and harmonious appearance of the historic building's revitalized roofing, the use of 19th-century tiles makes it possible and, more importantly, will in the future enable specialists to identify and differentiate between original and contemporary elements. In the opinion of the authors, this approach also perfectly fulfills the authenticity document requirements of the Nara [26] for the preservation of the originality of historical elements.

5. Conclusions

In the opinion of the authors, the use of tiles formed by hand and fired according to 19th-century technology in coal-fired Hoffman kilns not only meets the technical and conservation requirements in the case of historic buildings, but also prevents many of the disadvantages that accompany the replacement of roofing with new, machine-made contemporary roofing. The tiles, which are a precise reproduction of the geometry, texture, and color of the original roofing elements, are in keeping with the conservation doctrine of preserving the traces of our ancestors' activities and presence as accurately as possible. They do not introduce cognitive dissonance when viewing the monument, as is the case with additions or the replacement of original elements with completely new ones, but only supplement or finally replace the destroyed elements in a way that allows the monument to be fully perceived in its original form without falsifying its history.

There are doubts about the cost of making new handmade tiles or maintaining old tiles compared to using contemporary materials. The question is: Is it easier and quicker to replace the roof covering with a new one made of commonly available materials? While preserving for posterity the authenticity of heritage buildings, together with their scientific and historical significance, and fulfilling the requirements of conservation doctrine, we should revitalize the roof coverings of historic buildings in a way that allows the building's unadulterated past to be maintained. This choice should not be questioned.

Author Contributions: Conceptualization, K.A. and Ł.B.; methodology, Ł.B.; software, K.A. and Ł.B., M.N.-A., M.K. and M.P.-M.; validation, M.N.-A. and M.P.-M.; formal analysis, Ł.B. and M.P.-M.; investigation, K.A., Ł.B., M.N.-A. and M.P.-M.; resources, K.A., Ł.B., M.N.-A., M.P.-M. and M.K.; data curation, Ł.B. and M.P.-M.; writing—original draft preparation, K.A., Ł.B., M.N.-A., M.P.-M. and M.K.; writing—review and editing, K.A., Ł.B. and M.K.; visualization, Ł.B. and M.P.-M.; supervision, K.A. and Ł.B.; project administration Ł.B.; funding acquisition, K.A. All authors have read and agreed to the published version of the manuscript.

Funding: This research received no external funding.

Institutional Review Board Statement: Not applicable.

Informed Consent Statement: Not applicable.

Data Availability Statement: Not applicable.

Conflicts of Interest: The authors declare no conflict of interest.

References

1. Journal of Polish Laws 2003 No. 162 Item. 1568 Day 23 July 2003 on the Protection and Guardianship of Monuments. Available online: https://isap.sejm.gov.pl/isap.nsf/download.xsp/WDU20031621568/U/D20031568Lj.pdf (accessed on 5 November 2022).
2. *Conservation Officer's Handbook International Standards in Cultural Heritage Protection*; ICOMOS—Polish Committee of International Council on Monuments and Sites: Warsaw, Poland, 2015.
3. International Charter for the Conservation and Restoration of Monuments and Sites (The Venice Charter). In Proceedings of the 2nd International Congress of Architects and Technicians of Historic Monuments, Venice, Italy, 25–31 May 1964.
4. Ałykow, K.; Napiórkowska-Ałykow, M. On the inadequate modelling of the structure of architectural heritage. *J. Herit. Conserv.* **2015**, *41*, 59–64.
5. Ałykow, K.; Napiórkowska-Ałykow, M. Re-modeling of the Monumental Building According to Existing Standards on Cultural Heritage Protection. In *Structural Analysis of Historical Constructions*; Aguilar, R., Torrealva, D., Moreira, S., Pando, M.A., Ramos, L.F., Eds.; Springer: Cham, Switzerland, 2019. [CrossRef]
6. Łapacz, P.; Winkowski, M. The renovation of the Salt House in Luban. In *Quantitative and Qualitative Analysis of Historic Roof Tiles*; PKN: Wrocław, Poland, 2016.
7. *EN 538:1999*; Clay Roofing Tiles for Discontinuous Laying—Flexural Strength Test. PKN: Warszawa, Poland, 1999.
8. *EN 539-2:2013-07*; Clay Roofing Tiles for Discontinuous Laying—Determination of Physical Characteristics—Part 2: Test for Frost Resistance. PKN: Warszawa, Poland, 2013.
9. *EN 1304:2013-10*; Clay Roofing Tiles and Fittings—Product Definitions and Specifications. PKN: Warszawa, Poland, 2013.
10. *EN 539-1:2007*; Clay Roofing Tiles for Discontinuous Laying—Determination of Physical Characteristics—Part 1: Impermeability Test. PKN: Warszawa, Poland, 2007.
11. Krzywka, Ł. *Historical-Architectural Analysis and Conservation Proposals for the Salt House in Luban*; Salt House: Lubań, Poland, 2017.
12. Available online: https://medievalheritage.eu/pl/strona-glowna/zabytki/polska/luban-dom-solny/ (accessed on 10 August 2022).
13. Fokt, K.; Tekiela, Ł.; Bena, W.; Koreś, D. *Vademecum of Upper Łużyce History*; Stowarzyszenie Południowo-Zachodnie Forum Samorządu Terytorialnego "Pogranicze": Luban, Poland, 2010.
14. Tekiela, Ł. *Defence Walls*; Salt House: Lubań, Poland, 2014.
15. Gąsior, M. *Programme of Conservation and Restoration Work for the Salt House Building in Luban*; PKN: Wrocław, Poland, 2018.
16. Bajno, D.; Bednarz, Ł.; Nowak, T. Problems relating to assessment, repair and restoration of wooden roof structures in historic buildings, as exemplified by two case studies in southern Poland. *Adv. Mater. Res.* **2013**, *778*, 888–894.
17. Bednarz, L.; Bajno, D.; Matkowski, Z.; Skrzypczak, I.; Leśniak, A. Elements of Pathway for Quick and Reliable Health Monitoring of Concrete Behavior in Cable Post-Tensioned Concrete Girders. *Materials* **2021**, *14*, 1503. [CrossRef] [PubMed]
18. Kuśnierz-Krupa, D. Protection issues in selected European historic towns and their contemporary development. In *E3S Web of Conferences*; EDP Sciences: Ulis, France, 2018; p. 00043.
19. Chernyshev, D.; Ivashko, Y.; Kuśnierz-Krupa, D.; Dmytrenko, A. Role of natural landscape in perception of Ukrainian sacral architecture monuments. Landscape architecture. *Sci. J. Latv. Univ. Agric.* **2020**, *17*, 13–21.
20. Québec Declaration on The Preservation of The Spirit of Place (Genius Loci). In Proceedings of the ICOMOS 16th General Assembly, Quebec, QC, Canada, 30 September–4 October 2008. Available online: https://whc.unesco.org/uploads/activities/documents/activity-646-2.pdf (accessed on 5 November 2022).
21. Wang, X.; Wu, C.; Lu, Y.; Tian, M. The Synergy of Metadata and Metamodel through Algorithm Modeling—Case Study of the Roof Tiles in Yangxindian Palace (Beijing, China). *Appl. Sci.* **2022**, *12*, 7031. [CrossRef]
22. Basiricò, T.; Enea, D. Seismic and Energy Retrofit of the Historic Urban Fabric of Enna (Italy). *Sustainability* **2018**, *10*, 1138. [CrossRef]
23. Shen, J.; Li, L.; Wang, J.-P.; Li, X.; Zhang, D.; Ji, J.; Luan, J.-Y. Architectural Glazed Tiles Used in Ancient Chinese Screen Walls (15th–18th Century AD): Ceramic Technology, Decay Process and Conservation. *Materials* **2021**, *14*, 7146. [CrossRef] [PubMed]
24. Małachowicz, M. *900 Years of Ceramic Roof Tiles in Poland*; Oficyna Wydawnicza ATUT—Wrocławskie Wydawnictwo Oświatowe: Wrocław, Poland, 2022.
25. Recommendation on The Historic Urban Landscape. In Proceedings of the General Conference of UNESCO, Paris, France, 10 November 2011. Available online: https://whc.unesco.org/uploads/activities/documents/activity-638-98.pdf (accessed on 5 November 2022).
26. The Nara Document on Authenticity. In Proceedings of the Nara Conference on Authenticity, Nara, Japan, 1–6 November 1994. Available online: https://whc.unesco.org/document/116018 (accessed on 5 November 2022).

materials

MDPI

Article

Diagnostics of the RC Roofing Structure of the 100-Year-Old Municipal Theatre Facility

Marta Kosior-Kazberuk *, Janusz Ryszard Krentowski and Maciej Wardach

Faculty of Civil Engineering and Environmental Sciences, Bialystok University of Technology, Wiejska 45E, 15-351 Bialystok, Poland
* Correspondence: m.kosior@pb.edu.pl

Abstract: Reinforced concrete has been a widely used material for the construction of buildings for many decades. However, with the passage of time, the material characteristics and connection of structural elements gradually degrade. Development in measurement technology makes it possible to efficiently obtain data on the current state of the structure and material characteristics using non-destructive methods, with limited or no destructive testing. The paper presents the analysis of the condition of the reinforced concrete roof of a 100-year-old theatre building in terms of its further use after planned modernisation. The tests carried out in situ as well as the computational analysis of structure are described. Based on the test results, the current load-bearing capacity was assessed and the limit state conditions were checked. Limitations on the accuracy of the non-destructive test results in relation to the destructive test results were formulated. Options for the strengthening reinforced concrete beams with regard to structural and technological considerations were analysed.

Keywords: RC beams; NDT; historic structures; degradation; strengthening

Citation: Kosior-Kazberuk, M.; Krentowski, J.R.; Wardach, M. Diagnostics of the RC Roofing Structure of the 100-Year-Old Municipal Theatre Facility. *Materials* **2022**, *15*, 7438. https://doi.org/10.3390/ma15217438

Academic Editor: Baoguo Han

Received: 26 September 2022
Accepted: 21 October 2022
Published: 23 October 2022

1. Introduction

Poor management and lack of maintenance of buildings that have been in continuous use for more than a century threaten the durability of the structure and, consequently, the safety of users [1,2]. Ignoring the need for retrofit work can accelerate the degradation processes of building materials [3]. External structural elements, exposed to the direct effects of the environmental influence of the weather, will not wear out quickly if the facility is diagnosed correctly, and the repairs carried out can effectively extend the service life [4,5]. Degradation processes are influenced by both environmental factors and the natural 'ageing' of materials. The errors occurring at the design and construction stage, or incorrect operation or condition assessment, are also factors that can generate hazardous situations [6–9]. The public buildings should be constantly monitored and should comply with the most restrictive standards. Properly carried-out renovation work allows them to operate safely and regain their architectural qualities [10,11].

Reliability of structures, i.e., the ability to meet the requirements of load-bearing capacity, serviceability and durability, is a fundamental design issue. Methodological principles for the design of structures in Europe are included in the standard [12], which is based on the method of limit states and partial factors. In the case of calculation of newly designed structures, the materials with strength parameters adequate for the load-carrying requirements placed on them are adopted. The problem occurs with existing structures, for which static calculations must be carried out that take into account the actual physical and strength properties of materials and state of degradation. Before proceeding with design work for the reconstruction, renovation or strengthening of such structures, it is necessary to carry out a number of tests of material parameters. To avoid design errors, it is reasonable to analyse the cases of structural failures that have occurred. These failures occur in both industrial [13–15] and civil structures [16]. Analysis of the origin of unintentional structural failures provides a valuable testing ground for scientists and designers. The combination of advanced research methods together with numerical calculations allows the causes of loss of load-bearing capacity to be identified, which contributes to improving design work and avoiding the duplication of errors.

Public buildings constructed in the 20th century were usually constructed as masonry. Widely known ways of maintaining, repairing or strengthening such structures have been described in the literature [17,18]. However, between the wars, i.e., in the 1930s, reinforced concrete structures and mixed or combined structures such as reinforced concrete and masonry or reinforced concrete and steel structures were also built. These structures were usually characterised by a long service life, but nowadays, due to their service life of about a century, they require immediate repairs or complex modernisations. A change in the function of or way of using the building requires a detailed analysis of the conditions of the elements of historic structure and the properties of materials used over many years.

The development of measurement technologies is positively influencing the diagnosis of building structures. Modern testing methods such as digital image correlation [19–21], as well as laser scanning [22,23] and ultrasonic methods [24,25], allow for non-invasive monitoring of structures. Non-destructive testing of concrete elements has been popularised in [26–28]. Due to the exposure of the elements to aggressive industrial influences, chemical tests of the concrete also provide reliable results [29]. The extensive descriptions of experimental methods of detecting structural damage that use ultrasonic methods based on the nonlinear Lambda wave principle were presented in [30] and the methods based on colinear nonlinear mixed-frequency ultrasound can be found in [31]. The systematisation and summarisation of the latest methodologies and technologies for vibration-based structural health monitoring can be found in [32].

As a result of the conducted research, it becomes possible to carry out an analysis of the current technical state of existing structural and architectural solutions. The results of research also make it possible to carry out calculations that allow for an unambiguous assessment of the structure's load-bearing capacity [33]. Strengthened structural components are also being experimentally studied in research labs [34,35], contributing to the development of repair methods. In the case of complex modernisation of the building, all structural elements should comply with the limit state conditions specified in valid standards and regulations.

2. Historic Theatre Roof Structure

The described diagnostic measures are illustrated with an example of the research procedure and concepts for reinforcing the RC roof of historic theatre, located in north-eastern Poland. The main part of the theatre was built between 1933 and 1938. After nearly 100 years of exploitation, the city authorities decided that the building, which is listed as an historical monument, needed to be renovated. The aim of the planned work was to bring the facility up to current standards in terms of stage technology and acoustic requirements, and, above all, structural solutions to ensure the safety of users.

The load-bearing structure of the building consists of solid ceramic brick walls on lime mortar, based on concrete foundation walls. The inter-storey floors were built as reinforced concrete slabs supported on reinforced concrete beams and external walls. The roof slab was made as a monolithic reinforced concrete structure. The main structural elements of the roof are the reinforced concrete beams, varying in height from 1.00 m to 1.30 m, on which the box section roof is supported. The beams, with a clear span of 15.74 m, are supported on reinforced concrete columns with a cross-section of 0.50 × 0.70 m. Bevels were made in the support zones of the beams to stiffen the joints. The schematic view of the beams on the theatre plan is shown in Figure 1a. The structural cross-sections are presented in Figure 1b–d.

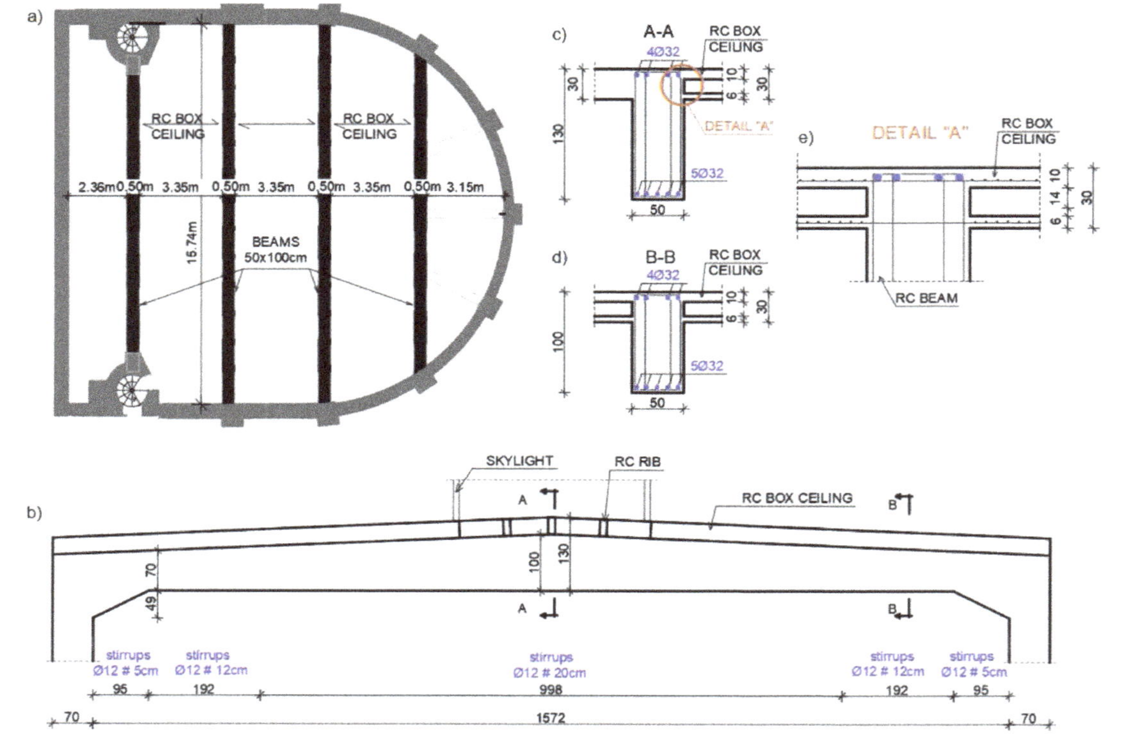

Figure 1. The layout of the roof ceiling: (**a**) plan of the beams; (**b**) view of the beam; (**c**,**d**) structural cross-sections of the beam; (**e**) layout of the box ceiling.

The box section ceiling structure was made as a monolithic one (Figure 2b). The dimensions of the ceiling are shown in Figure 1e. In the central part of the ceiling, between two central beams, there is a skylight of a width corresponding to the ceiling span between the beams. The main structural elements of the ceiling in this zone are the ribs (Figure 2a) with cross section 0.30 × 0.12 m. The axial spacing of the ribs is approximately 0.90 m. The ceiling is insulated with hard mineral wool and covered with two layers of roofing felt.

(**a**) (**b**) (**c**)

Figure 2. View of the beams: (**a**) at the skylight location; (**b**) near the column support; (**c**) connection to the column.

A steel-framed technological platform was suspended to the beams above the stage (Figure 3), providing additional load to the existing structure. The main structural elements are steel hangers, to which beams made of 180 mm high rolled I-beams are attached. C100 joists (Figure 3a) are supported on the beams, with steel deck gratings made of flat bars. Lighting bridges, winches and other equipment necessary for the theatre's stage are mounted on the technical platform.

(a) (b) (c)

Figure 3. View of the technological platform: (**a**) winches: (**b**) the platform grid: (**c**) elements attached to the lower reinforcement of the beams.

Due to the planned change, i.e., increased loads resulting from the necessity to use new technology, as well as new acoustic requirements, a variant analysis of how to adapt the structure to current regulations became necessary. Whole equipment comprising the so-called stage technology was designed to be suspended from the reinforced concrete roof. A similar solution was applied in the theatre building to that time. An important design aspect is also the current fire protection standards, which are more stringent than years ago. The main load-bearing elements, especially the beams, are vulnerable to damage during a fire. Elements weakened by damage to the cover layer and corrosion of the reinforcing bars are less able to withstand the temperature of a fire, which can consequently lead to the disaster of the whole structure [36]. Analysis of the real state of the structure, additionally taking into account the thermal effects on the state of damaged structural elements, recommends that specialised calculation models are used [37–39]. The effectiveness of the obtained computational results is determined by the quality of the diagnostic data entered into the calculation programs.

3. Test Procedures for Reinforced Concrete Beams

To determine the actual strength parameters, as well as the progress of the degradation processes of the structural elements, the series of in situ tests were carried out. Non-destructive (Figure 4) and visual methods of concrete and steel testing were used, and local excavations were made. Due to lack of knowledge regarding the condition of the structure at current stage of the study, it was decided not to take specimens for destructive testing of concrete and steel. The results of non-destructive testing are subjected to a higher risk of error in comparison to the results obtained by destructive methods. However, in many cases, they are the only source of data allowing a preliminary assessment of the technical condition of structure.

(**a**) (**b**) (**c**)

Figure 4. Non-destructive testing: (**a**) Pulse-Echo method; (**b**) ultrasonic testing using longitudinal wave transducers: (**c**) concrete moisture testing.

Visual examination revealed that repair work to the beams consisting of reprofiling was conducted in the past (Figure 5a). The reprofiling had been carried out incorrectly, as the concrete cover of the reinforcing bars was found to be missing on significant areas of beam surface. In addition, the repair mortar easily peeled off after a light impact. This means that the protective layer did not achieve proper adhesion with the structural concrete. Numerous irregular cracks, with an opening width that did not exceed 0.5 mm, were found on the vertical surfaces of beams (Figure 5b,c).

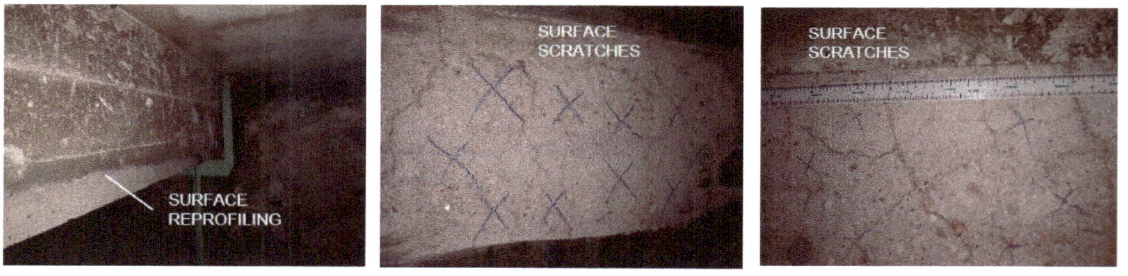

(**a**) (**b**) (**c**)

Figure 5. Visual tests: (**a**) view of reprofiled surfaces; (**b**,**c**) surface cracks.

After removing the roughcast and outer layer of concrete, it was possible to determine that the cracks in the area of the opencast were superficial. No excessive cracks were found in the span zone or in the support zone as a result of exceeding the bending moment limits or the shear force. Potential scratches in these zones were covered up by the applied cement mortar repair treatments. No excessive scratches were found on the reprofiled lower surfaces of beams. The roof surface was airtight, and no signs of streaking were observed. The test of moisture in the structural materials, carried out using a digital moisture meter (Figure 4c), gave the results ranging from 45 to 60% RH.

To estimate the compressive strength and homogeneity of the concrete of several structural elements, a sclerometric method was used. On both sides of all beams, 5 surfaces each were prepared with power tools to read the number of rebounds. On this basis, the strength of concrete corresponding to contemporary C16/20 class was estimated, noting that the concrete is strongly heterogeneous. Nowadays, concrete of this class is not applied for the main structural elements, but only for secondary elements or base layers. A dull sound was found locally, which is characteristic of delaminated elements with voids and discontinuities inside the concrete section. The varying rebound number can also indicate carbonisation of the concrete, which leads to hardening of its surface layer. The process of carbonisation is associated with the loss of passivation of reinforcing steel, which can result in the loss of adhesion of the reinforcement and its corrosion.

In order to confirm the results of the sclerometric tests, the ultrasonic tests were carried out. Two-sided access to the component using longitudinal wave transducers was used (Figure 4a). Six measurements were taken on each beam—2 readings in each of the support and the span zones. Regardless of the location of the transducers, it was not possible to carry out a measurement due to the high interference and wave dispersion, confirming the strong degradation of the reinforced concrete elements. Locally, a surface wave velocity of about 2200 m/s was measured, which, according to [40], qualifies the concrete as weak. Next, measurements were conducted using transverse waves with a Pulse-Echo transducer (Figure 4b). Readings were taken for both sides of the beams where measurements were made with longitudinal wave transducers. The ultrasonic wave velocity was extremely low and averaged 1100 m/s, where the typical value for concrete is 2000–2200 m/s. In some places, the measures were impossible. The results confirmed the hypothesis of discontinuity of the medium due to internal defects in the concrete structure. Plots of longitudinal wave propagation (measured with single-sided access) and transverse wave propagation (using the Pulse-Echo head) are presented in Figure 6.

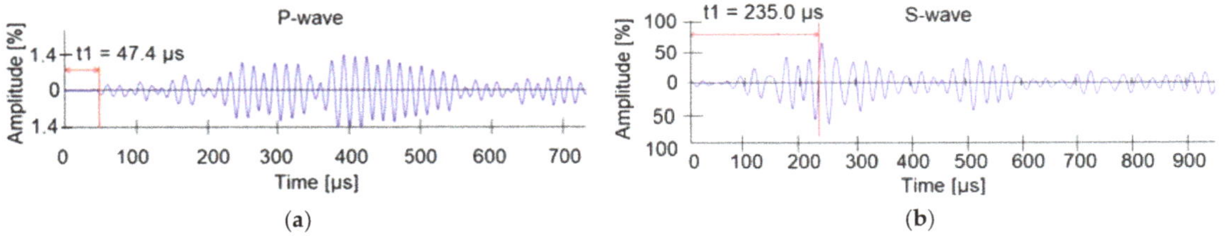

Figure 6. Wave propagation diagram: (**a**) longitudinal wave; (**b**) transverse wave.

Electromagnetic scanning of the beams was carried out to determine the quality of the reinforcement work (Figure 7). The measurements made it possible to assess the distribution of the reinforcement, which was then implemented in the static calculations. Insufficient cover thicknesses were locally found. Initial corrosion processes of the steel reinforcement were inventoried at these locations. The too-low cover thickness or local cover losses have a negative effect on the structure's resistance to fire temperatures.

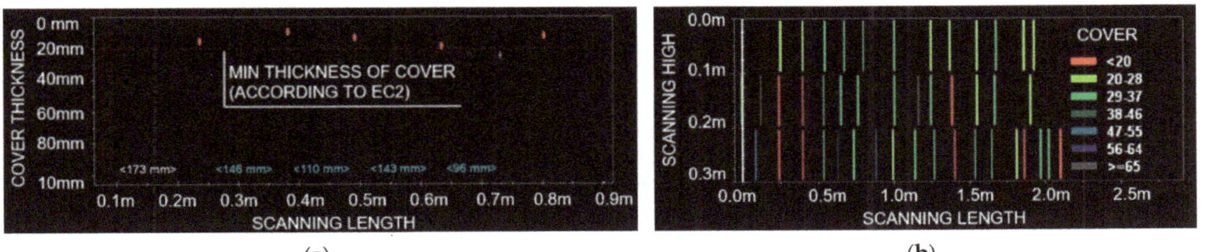

Figure 7. Results of electromagnetic scans: (**a**) linear scan; (**b**) multilinear scan.

The diameter and distribution of the reinforcement were determined on the examination of excavations, which were carried out on the beam surface and the analysis of areas without concrete cover (Figure 8). The number of excavations was limited due to concerns about the poor technical condition of beams.

(a) (b) (c)

Figure 8. Excavations on beam's surface: (**a**) at the point where the main reinforcement bars are connected to the winch; (**b**) at the point of insufficient cover in the beam span; (**c**) at the point of insufficient cover in the bevel.

The research allowed it to be established that the bottom reinforcement of the beam was made up of 5 Ø32 mm diameter bars of non-ribbed steel. The transverse reinforcement was made of four-cut stirrups of Ø12 mm bars of non-ribbed steel. In order to simplify the implementation of the scan results into the numerical model, the distances between the stirrups were averaged due to the very uneven spacing of the stirrups. At the supports, the average spacing of the stirrups was assumed to be every 5 cm, further on every 12 cm and in the middle of the span every 20 cm. The box ceiling ribs were reinforced with 2 Ø12 bars at the bottom. The stirrups were made as double-cut Ø8 mm bars and were spaced at 15 cm and 20 cm intervals. Based on archival data from buildings constructed at the same time and constructed in the same area as the building under investigation, the yield strength of the reinforcement bars was assumed to be 230 MPa. Details of the beam reinforcement are shown in Figure 1c–e.

4. Computational Verification of Stress State

The calculations were performed in ANSYS [41], assuming the stress–strain relationships of concrete and steel according to EC2 [42] Section 3.1.7 and Section 3.2.4. The FEM mesh was assumed to be rectangular in shape, where the maximum side dimension does not exceed 250 mm. The grid size results from the most favourable calculation time with a satisfactory level of result accuracy. Beams and columns were modelled with SOLID186 elements, i.e., solid elements defined by 20 nodes with three degrees of freedom per node. The element is characterised by ductility, high deflection and large deformation capacity. Reinforcing bars were modelled as reinforcement of the solid elements using REINF264.

On the basis of sclerometric tests, the strength of the concrete was estimated to be equivalent to contemporary class C16/20 (f_{ck} = 16 MPa, f_{cd} = 11.43 MPa), and the yield strength of the reinforcing steel was assumed to be f_{yk} = 230 MPa (f_{yd} = 200 MPa). The number and spacing of bars were assumed on the basis of the inventory, the excavations and the results of the electromagnetic tests. Static calculations for the design values of the strength parameters of concrete and steel were carried out according to [42]. The effective width of the box section floor, representing the top shelf of the beam, was also taken into collaboration. The calculation of the effective width of the b_{eff} was carried out according to [42].

The calculations were carried out for two static schemes, i.e., a frame scheme with the beams rigidly connected to the columns and an articulated scheme with the simply supported beams. The plasticisation of the nodes can lead to articulation and hence a change in the static scheme from a frame to a simply supported beam. In view of the lack of reliable data in terms of the formation of frame joints, the uncertain quality of the concrete and steel, and the expected service life over the next few decades, it is reasonable to carry out calculations for both variants.

The beams were loaded with dead loads of the roof (g = 5.6 kN/m^2), as well as climatic loads, i.e., wind (w = 0.35 kN/m^2) and snow (s = 1.28 kN/m^2), and service loads (q = 2.0 kN/m^2). Loads were assumed in accordance with standards [43–45], while calculation combinations were formulated in accordance with [12]. The results of the calculation are shown in Figure 9. The programme automatically takes into account the dead weight of the beam, which is approximately 14.38 kN/m and represents about 40% of the total dead loads.

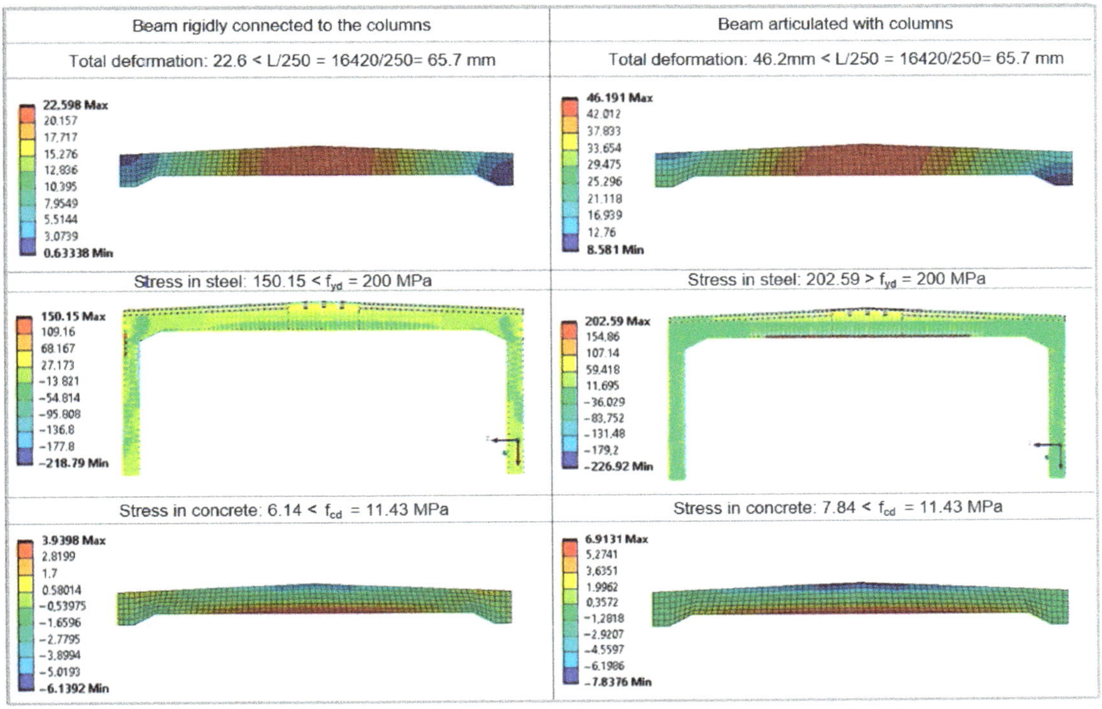

Figure 9. Static calculation results.

On the basis of the calculations, it was indicated that the ultimate limit state and serviceability limit state would not be exceeded for the beams rigidly connected to the columns. In the case of the simply supported beam scheme, the stresses of 202.6 MPa in the reinforcing steel exceed the normal values of 200 MPa and the ultimate limit state is not fulfilled by only 1%. It is worth noting that in both static schemes the serviceability limit state due to deflection is fulfilled.

Especially dangerous for the operation of the beams will be a change in the static scheme with plasticisation of the joints connecting the columns with the beam. This situation will result in a redistribution of forces in the beam and an increase in the bending moment in the span zone. In this situation, the bending capacity of the beam will be exceeded, which poses a safety risk to the structure and the users of the building. This phenomenon may occur, for example, in the case of locally lower strength parameters of materials used to construct the beams compared to the assumed design material features. These lower material parameters may by caused by poorly compacted concrete mix resulting in concrete heterogeneity, delamination and cracking.

However, it should be noted that the static-strength analysis was carried out without taking into account reductions in the load-bearing capacity of the beams caused by internal defects such as possible poor compaction of the concrete mix, delaminations, concrete cracking and ageing. As a result, there is a real risk of a higher strain on the elements due to the very poor quality of the concrete and the reinforcement work. The calculations carried out for the two static schemes showed that in the case of plasticisation of the joints, the ultimate limit state would be exceeded, even without taking into account internal defects. Therefore, in the case under analysis, consideration of the decrease in bearing capacity caused by such degradation would be negligible.

5. Analysis of Opportunities for Strengthening

With regard to the risk of exceeding the ultimate limit state and serviceability limit state as a result of new designed loads, it is necessary to select an economically and constructionally optimal design solution. The problem of strengthening reinforced concrete beams has been analysed many times in available monographs and scientific articles [46–49]. Based on published solutions and their own research experience, the authors have made a variant assessment of the possibility of reinforcing the existing structure in terms of its continued safe use.

5.1. Concretising the Existing Structure

One option for increasing the load-bearing capacity is to concretise the structural element [50], i.e., to increase its dimensions, above all, the useful depth of reinforced concrete section. This method simultaneously rebuilds the lost cover layer. During the selection of the sprayed mix, the factors of environmental aggression and the expected strength parameters have to be considered. In addition to the parameters of sprayed concrete, the quality of the strengthening is also affected by the parameters of shotcrete and the type of machinery used [51]. These factors generate a problem in determining the parameters of the applied concrete. The literature [52,53] presents the results of experimental studies related to the determination of Young's modulus and shear capacity. Thus, this method is relatively simple in terms of execution, but the reinforcement parameters are very difficult to determine precisely. In the case of the analysed beams, for which multiple reprofiling had previously been used, concretising the element would have been troublesome due to the variable adhesion of the individual concrete layers, i.e., the original and after reprofiling. The reinforced concrete must also have an adequate compressive strength, which is very low in the case of the beams analysed. The application of such a method requires the removal of the carbonatised concrete layer up to the surface of the material with correct elastic and strength characteristics. On the basis of preliminary destructive tests, it was determined that it would be necessary to remove the concrete to a depth of several centimetres around the entire circumference of the existing beams. The problem of connecting the reinforcing layer to the existing uneven-surfaced structure could be solved by using additional anchors made of ribbed steel bars, set to a depth of several dozens of centimetres, through high-strength mortar.

The authors also considered the option of surrounding the cleaned surfaces of beams with 3 mm diameter bar nets, fixed with bars embedded permanently in the original structure. A few centimetres' layer of sprayed concrete could be effectively placed on such prepared surfaces. The similar solution was described in [54], where the supporting structure of industrial tanks was reinforced.

5.2. Demolition of the Roof Structure

The possible option is to remove the slab of box section roof while leaving the beams. In this case the beams relief will be provided but, at the same time, the beams will lose their plate bracing. The partial reduction in load bearing capacity can be expected as the top shelf of the T-section of beam will be removed. Between the beams, lightweight steel trusses can be constructed on which the new roof, made of structural steel plate and a purlin system of cold-formed steel sections, would be supported. The steel trusses would be supported on steel replacements articulated on the existing columns. This way, it would be possible to relieve the load off the existing beams and to transfer all the loads to the new structure. The beams, carrying only their own weight in this option, could be used to carry process loads, e.g., from light installations.

The most time-consuming and expensive solution is to remove the entire reinforced concrete ceiling with beams and build a new roof in lightweight steel construction with trapezoidal sheet metal cladding. However, acoustic considerations are a contraindication to the implementation of a steel-framed roof. Due to the low weight of such a structure and the specific characteristics of the steel elements, additional screens will be required to dampen vibrations caused by acoustic waves. In case of a decision to demolish the beams, the most favourable solution in terms of lead time would be to use prefabricated prestressed beams. Once the existing elements had been cut out, the columns would need to be profiled to accommodate the new reinforced concrete girders.

It should be noted that the number of box ceiling tests are also required for both the above solutions. Aging processes and the influence of the external environment (the roof is most exposed to water and temperature changes) may have reduced its load-bearing capacity, just as in the case of beams. Determining the degradation state of the roof should precede design work for demolition. It is important to prevent structural failure also in the last stage of the structure's life, i.e., its dismantling. Improperly planned demolition work can lead to a structural disaster and risk the lives of workers.

5.3. Other Strengthening Concepts

Another strengthening option would be to place steel trusses at mid-span of the beams, without removing the box section roof. The trusses would be supported on the columns by steel or reinforced concrete replacements. In this way, the beams would be relieved of half the applied loads. The risk is the change in the static scheme of the floor slab resulting from the addition of more supports. The execution problem is the need to weld a steel structure of low stiffness to a reinforced concrete section in order to get the existing floor to cooperate with the new truss.

Strengthening by means of reinforcement with composite tapes is often used to strengthen reinforced concrete beams [47,55]. Such strengthened structural elements are characterised by various types of damage, which are widely described in [49]. Predictive models for typical damage are also known [56]. The creep of the concrete and the type of adhesion are important in terms of the reliability of such a strengthening [57]. Various configurations of strengthening reinforced concrete beams with composites were experimentally tested in [58]. The commonly used methods of increasing the load capacity can be divided into passive and active ones. The passive method consists of passively reinforcing the tension zone of the component with tapes. The composite tape is active only when the deformation of the structure has increased. In the case of beams considered, this method would be ineffective, as the dominant load is the dead weight of the roof and beams. The active method relies on the external pre-stressing of the structure by means of laminations using tension devices attached to the members to be strengthened. The undoubted advantage of this method is that the structure does not need to be unloaded but the problem is the making of the anchor block fixings. Due to uncertain quality and strong heterogeneity of the concrete, there is a risk that the attachment of the blocks to the reinforced concrete beams will not be carried out correctly. In addition, due to the close spacing of the reinforcing bars, it is impossible to fix the anchor bars without damaging the reinforcement. For both methods, restrictive fire regulations are also a contraindication to their application. The elements would have to be secured in accordance with current standards, which would entail additional costs as well as additional incremental loads.

Nowadays, with the development of construction chemistry products, many modern, specialised mortars are used for dry or wet spray application. The papers [59,60] present the results of experimental studies of strengthening using repair mortars. The mixtures are characterised by the possibility of application in any exposure class of structural elements. Properly selected mixtures can be used even in the most aggressive chloride ion environment, i.e., in coastal engineering [61]. The considered beams had previously been reprofiled, but the repair work was carried out incorrectly. The placement of another spray layer is considered to be a technically incorrect solution, which will not fulfil the expected strengthening effect due to the lack of cooperation with the original reinforced concrete core and will not provide the required load-bearing capacity for the newly designed loads.

6. Discussion

Historic structures in operation should be strengthened in such a way that changes occurring during the design life, taking into account environmental influences and the expected level of maintenance, do not reduce the performance of the structure below the intended level. The design life of the theater's continued use, according to [12], is 100 years. However, the poor quality of concrete and reinforcement work, as well as numerous defects in the structure of the concrete and inadequate reprofiling, indicate the unquantifiable risk of using the beams for further service [8].

The observed cracks on the lateral surfaces of the beams are mainly caused by concrete shrinkage. The size of the crack is significantly influenced by the large thickness of the reinforcement lagging, as well as the insufficient amount and spacing of the horizontal shrinkage reinforcement. Cracks of this type do not threaten the safety of the structure but affect the homogeneity of concrete.

The properties of steel bars produced in the early 20th century may now differ significantly from their original properties, as they have been subject to an ageing process. As steel ages, its tensile strength, yield point and hardness change. The loss of steel's ductile properties can result in unsignalised damage or failure of the beams as a result of exceeding the yield strength. It should be taken into account that in the analysed beams, the ageing process of the steel will continue to progress. Heterogeneity and progressive carbonation of concrete can negatively affect the properties in terms of protection of reinforcing bars against corrosion. Progressive corrosion, on the other hand, is associated with the loss of bar cross-sections, which corresponds to a reduction in load-bearing capacity.

The listed defects may cause a local reduction in the strength parameters of the materials resulting in the risk of exceeding the ultimate limit state. The formation of articulations in the connection zones between columns and beams will cause an increase in the bending moment in the beam spans, resulting in a risk of failure or structural catastrophe.

7. Conclusions

The research on the historic ceiling and the discussion of reinforcement options allowed for the formulation of the following conclusions:

- in the case of historic unit objects, limited access to structural elements and the inability to conduct full destructive testing force the use of a combination of non-destructive testing methods, without significant interference with the historic building;
- a properly selected combination of non-destructive testing methods allowed for a comprehensive assessment of the technical condition of the facility, necessary to develop variants of its strengthening;
- in the case that we are faced with the dilemma of how to preserve a monument in accordance with the conservation doctrine and at the same time secure it in such a way that it meets the requirements of the safe stay of people, it is necessary to analyse the options for repairing the object;
- in the case of historic buildings, the amount of questionable information and assumptions is so large that both the survey and numerical calculations only estimate the results. Therefore, the experience of a construction appraiser who makes proper design decisions is necessary. According to the principle of "Practice Makes Perfect," the greater the experience of the appraiser, the greater the certainty regarding the obtained results of research and numerical analysis. This affects the effectiveness in making decisions on the continued exploitation of the building.

Author Contributions: Conceptualization, M.K.-K., J.R.K. and M.W.; methodology, M.K.-K. and J.R.K.; formal analysis, J.R.K.; investigation, M.K.-K. and M.W.; resources, J.R.K.; writing—original draft preparation, M.W.; writing—review and editing, M.K.-K. and J.R.K.; visualisation, M.K.-K. and M.W.; supervision, J.R.K.; project administration, M.K.-K.; funding acquisition, M.K.-K. All authors have read and agreed to the published version of the manuscript.

Funding: The research was carried out within the scope of work no. WZ/WB-IIL/2/2020 and no. WI/WB-IIL/2/2021 and financed from the resources for science of Ministry of Education and Science of Poland.

Institutional Review Board Statement: Not applicable.

Informed Consent Statement: Not applicable.

Data Availability Statement: The data presented in this study are available on request from the corresponding author.

Conflicts of Interest: The authors declare no conflict of interest.

References

1. Drobiec, Ł.; Grzyb, K.; Zając, J. Analysis of Reasons for the Structural Collapse of Historic Buildings. Sustainability. *Sustainability* **2021**, *13*, 10058. [CrossRef]
2. Berkowski, P.; Kosior-Kazberuk, M. Material and Structural Destruction of Concrete Elements in the Industrial Environment. *Procedia Eng.* **2017**, *172*, 96–103. [CrossRef]
3. Nowogońska, B. Consequences of Abandoning Renovation: Case Study-Neglected Industrial Heritage Building. *Sustainability* **2020**, *12*, 6441. [CrossRef]
4. Knyziak, P.; Kanoniczak, M. Difficulties in Operation of Elevations in Large-Panel Buildings. *IOP Conf. Ser. Mater. Sci. Eng.* **2019**, *661*, 012059. [CrossRef]
5. Berkowski, P.; Kosior-Kazberuk, M. Historical and structural aspects of the durability and maintenance of a reinforced concrete market hall building from the early 20th century. In *Concrete Solution, Proceedings of the 6th International Conference on Concrete Repair, Thessaloniki, Greece, 20–23 June 2016*; CRC Press: Boca Raton, FL, USA, 2016; pp. 335–343.
6. Krentowski, J.R. Materials Assessment of Destructive Impact of Different Factors on Concrete Structures Durability. *Materials* **2021**, *15*, 225. [CrossRef]
7. Krentowski, J.R.; Ziminski, K. Consequences of an Incorrect Assessment of a Structure Damaged by Explosion. *Eng. Fail. Anal.* **2019**, *101*, 135–144. [CrossRef]
8. Knyziak, P. The Impact of Construction Quality on the Safety of Prefabricated Multi-Family Dwellings. *Eng. Fail. Anal.* **2019**, *100*, 37–48. [CrossRef]
9. Knyziak, P. The Quality and Reliability in the Structural Design, Production, Execution and Maintenance of the Precast Residential Buildings in Poland in the Past and Now. *Key Eng. Materials* **2016**, *691*, 420–431. [CrossRef]
10. Gil-Mastalerczyk, J.; Gil, R. Difficult Geotechnical Conditions Under the Palace Complex, Case Study from Cianowice, Near Krakow, Poland. *IOP Conf. Ser. Earth Environ. Sci.* **2016**, *44*, 022031. [CrossRef]
11. Gil-Mastalerczyk, J. The Regained Beauty of the Pilgrimage Monastery in Imbramowice, on the 900th Anniversary of the Norbertine Order: Religious Heritage Challenges and Threats. *Wiadomości Konserw.* **2021**, *64*, 101–114.
12. *EN 1990*; Eurocode—Basis of Structural Design. The European Union: Brussels, Belgium, 2004.
13. Bahrami, A.; Taheri, P. A Study on the Failure of AISI 304 Stainless Steel Tubes in a Gas Heater Unit. *Metals* **2019**, *9*, 969. [CrossRef]
14. Bahrami, A.; Mousavi Anijdan, S.H.; Taheri, P.; Yazdan Mehr, M. Failure of AISI 304H Stainless Steel Elbows in a Heat Exchanger. *Eng. Fail. Anal.* **2018**, *90*, 397–403. [CrossRef]
15. Pourmohammad, H.; Bahrami, A.; Eslami, A.; Taghipour, M. Failure Investigation on a Radiant Tube in an Ethylene Cracking Unit. *Eng. Fail. Anal.* **2019**, *104*, 216–226. [CrossRef]
16. Augenti, N.; Parisi, F. Learning from Construction Failures Due to the 2009 L'Aquila, Italy, Earthquake. *J. Perform. Constr. Facil.* **2010**, *24*, 536–555. [CrossRef]
17. Carozzi, F.G.; Milani, G.; Poggi, C. Mechanical Properties and Numerical Modeling of Fabric Reinforced Cementitious Matrix (FRCM) Systems for Strengthening of Masonry Structures. *Compos. Struct.* **2014**, *107*, 711–725. [CrossRef]
18. de Santis, S.; de Felice, G. Steel Reinforced Grout Systems for the Strengthening of Masonry Structures. *Compos. Struct.* **2015**, *134*, 533–548. [CrossRef]
19. Golewski, G.L. Evaluation of Fracture Processes under Shear with the Use of DIC Technique in Fly Ash Concrete and Accurate Measurement of Crack Path Lengths with the Use of a New Crack Tip Tracking Method. *Measurement* **2021**, *181*, 109632. [CrossRef]
20. Golewski, G.L. Measurement of Fracture Mechanics Parameters of Concrete Containing Fly Ash Thanks to Use of Digital Image Correlation (DIC) Method. *Measurement* **2019**, *135*, 96–105. [CrossRef]
21. Krassowska, J.; Kosior-Kazberuk, M.; Berkowski, P. Shear Behavior of Two-Span Fiber Reinforced Concrete Beams. *Arch. Civ. Mech. Eng.* **2019**, *19*, 1442–1457. [CrossRef]

22. Pawłowicz, J. Digital Survey of Damages on the Façade of a Historical Building. *Acta Sci. Pol. Archit. Bud.* **2021**, *20*, 41–50. [CrossRef]
23. Pawłowicz, J.A. The TLS Technique as a Way of Identification and Measurement of Damaged Elements of a Historic Sacral Building. *MATEC Web Conf.* **2019**, *284*, 08007. [CrossRef]
24. Schabowicz, K. Non-Destructive Testing of Materials in Civil Engineering. *Materials* **2019**, *12*, 3237. [CrossRef] [PubMed]
25. Schabowicz, K. Materials Testing of Materials and Elements in Civil Engineering. *Materials* **2021**, *14*, 3412. [CrossRef]
26. Hoła, J.; Schabowicz, K. State-of-the-Art Non-Destructive Methods for Diagnostic Testing of Building Structures—Anticipated Development Trends. *Arch. Civ. Mech. Eng.* **2010**, *10*, 5–18. [CrossRef]
27. Hoła, J.; Bień, J.; Sadowski, L.; Schabowicz, K. Non-Destructive and Semi-Destructive Diagnostics of Concrete Structures in Assessment of Their Durability. *Bull. Pol. Acad. Sci. Tech. Sci.* **2015**, *63*, 87–96. [CrossRef]
28. Wardach, M.; Krentowski, J.R.; Knyziak, P. Degradation Analyses of Systemic Large-Panel Buildings Using Comparative Testing during Demolition. *Materials* **2022**, *15*, 3770. [CrossRef] [PubMed]
29. Hulimka, J.; Kału, M. Basic Chemical Tests of Concrete during the Assessment of Structure Suitability-Discussion on Selected Industrial Structures. *Appl. Sci.* **2020**, *10*, 358. [CrossRef]
30. Chen, H.; Liu, M.; Chen, Y.; Li, S.; Miao, Y. Nonlinear Lamb Wave for Structural Incipient Defect Detection with Sequential Probabilistic Ratio Test. *Secur. Commun. Netw.* **2022**, *2022*, 9851533. [CrossRef]
31. Chen, H.; Li, S. Collinear Nonlinear Mixed-Frequency Ultrasound with FEM and Experimental Method for Structural Health Prognosis. *Processes* **2022**, *10*, 656. [CrossRef]
32. Zhang, C.; Mousavi, A.A.; Masri, S.F.; Gholipour, G.; Yan, K.; Li, X. Vibration Feature Extraction Using Signal Processing Techniques for Structural Health Monitoring: A Review. *Mech. Syst. Signal Process.* **2022**, *177*, 109175. [CrossRef]
33. Berkowski, P.; Kosior-Kazberuk, M. Construction History as a Part of Assessment of Heritage Buildings. *Procedia Eng.* **2016**, *161*, 85–90. [CrossRef]
34. Ombres, L. Flexural Analysis of Reinforced Concrete Beams Strengthened with a Cement Based High Strength Composite Material. *Compos. Struct.* **2011**, *94*, 143–155. [CrossRef]
35. Huang, H.; Huang, M.; Zhang, W.; Guo, M.; Pospisil, S. Seismic Performance of Predamaged RC Columns Strengthened with HPFL and BSP under Combined Loadings. *Eng. Struct.* **2020**, *203*, 109871. [CrossRef]
36. Knyziak, P.; Kowalski, R.; Krentowski, J.R. Fire Damage of RC Slab Structure of a Shopping Center. *Eng. Fail. Anal.* **2019**, *97*, 53–60. [CrossRef]
37. Głowacki, M.; Abramowicz, M.; Kowalski, R. The Influence of Reinforcement Temperature on Stiffness of Reinforced Concrete Beams in Fire Conditions. *Bud. I Archit.* **2013**, *12*, 115–122. [CrossRef]
38. Kowalski, R. The Use of Eurocode Model of Reinforcing Steel Behavior at High Temperature for Calculation of Bars Elongation in RC Elements Subjected to Fire. *Procedia Eng.* **2017**, *193*, 27–34. [CrossRef]
39. Kowalski, R. Mechanical Properties of Concrete Subjected to High Temperature. *Archit. Civ. Eng. Environ.* **2010**, *3*, 61–70.
40. Dermawan, A.S.; Dewi, S.M.; Wisnumurti; Wibowo, A. Performance Evaluation and Crack Repair in Building Infrastructure. *IOP Conf. Ser. Earth Environ. Sci.* **2019**, *328*, 012007. [CrossRef]
41. *Ansys 2021 R1*; ANSYS Inc.: Canonsburg, PA, USA, 2021.
42. *EN 1992-1-1:2004*; Eurocode 2—Design of Concrete Structures. Part 1-1. General Rules and Rules for Buildings. The European Union: Brussels, Belgium, 2004.
43. *EN 1991-1-1:2004*; Eurocode 1:—Actions on Structures-Part1-1: General Actions-Densities, Self-Weight, Imposed Loads for Buildings. The European Union: Brussels, Belgium, 2004.
44. *EN 1991-1-4*; Eurocode 1: Actions on Structures—Wind Actions. The European Union: Brussels, Belgium, 2008.
45. *EN 1991-1-3*; Eurocode 1: Actions on Structures—Snow Loads. The European Union: Brussels, Belgium, 2005.
46. Raoof, S.M.; Koutas, L.N.; Bournas, D.A. Textile-Reinforced Mortar (TRM) versus Fibre-Reinforced Polymers (FRP) in Flexural Strengthening of RC Beams. *Constr. Build. Mater.* **2017**, *151*, 279–291. [CrossRef]
47. Kałuża, M.; Ajdukiewicz, A. Comparison of Behaviour of Concrete Beams with Passive and Active Strengthening by Means of CFRP Strips. *Archit. Civ. Eng. Environ.* **2008**, *1*, 51–64.
48. Bashandy, A.A. Flexural Strengthening of Reinforced Concrete Beams Using Valid Strengthening Techniques. *Arch. Civ. Eng.* **2013**, *59*, 275–293. [CrossRef]
49. Yao, J.; Teng, J.G. Plate End Debonding in FRP-Plated RC Beams—I: Experiments. *Eng. Struct.* **2007**, *29*, 2457–2471. [CrossRef]
50. Krainskyi, P.; Khmil, R.; Blikharskiy, Z. The Strength of Reinforced Concrete Columns, Strengthened by Reinforced Concrete Jacketing under Loading. *J. Civ. Eng. Environ. Archit.* **2015**, *XXXII*, 209–220. [CrossRef]
51. Ginouse, N.; Jolin, M. Mechanisms of Placement in Sprayed Concrete. *Tunn. Undergr. Space Technol.* **2016**, *58*, 177–185. [CrossRef]
52. Galobardes, I.; Cavalaro, S.H.; Aguado, A.; Garcia, T. Estimation of the Modulus of Elasticity for Sprayed Concrete. *Constr. Build. Mater.* **2014**, *53*, 48–58. [CrossRef]
53. Garcia, T.; Blanco, A.; Cavalaro, S.H.P. Shear Behaviour of Sprayed Concrete. *Constr. Build. Mater.* **2016**, *124*, 722–731. [CrossRef]
54. Krentowski, J.R.; Knyziak, P. Degradation and Reinforcement of Industrial Gas Tank Support Structures. Thirty-Year Long Monitoring. *IOP Conf. Ser. Mater. Sci. Eng.* **2017**, *245*, 032054. [CrossRef]
55. Guo, R.; Ren, Y.; Li, M.; Hu, P.; Du, M.; Zhang, R. Experimental Study on Flexural Shear Strengthening Effect on Low-Strength RC Beams by Using FRP Grid and ECC. *Eng. Struct.* **2021**, *227*, 111434. [CrossRef]

56. Teng, J.G.; Yao, J. Plate End Debonding in FRP-Plated RC Beams—II: Strength Model. *Eng. Struct.* **2007**, *29*, 2472–2486. [CrossRef]
57. Hamed, E.; Chang, Z.-T. Effect of Creep on the Edge Debonding Failure of FRP Strengthened RC Beams—A Theoretical and Experimental Study. *Compos. Sci. Technol.* **2013**, *74*, 186–193. [CrossRef]
58. Kotynia, R.; Oller, E.; Marí, A.; Kaszubska, M. Efficiency of Shear Strengthening of RC Beams with Externally Bonded FRP Materials—State-of-the-Art in the Experimental Tests. *Compos. Struct.* **2021**, *267*, 113891. [CrossRef]
59. Phoo-ngernkham, T.; Sata, V.; Hanjitsuwan, S.; Ridtirud, C.; Hatanaka, S.; Chindaprasirt, P. High Calcium Fly Ash Geopolymer Mortar Containing Portland Cement for Use as Repair Material. *Constr. Build. Mater.* **2015**, *98*, 482–488. [CrossRef]
60. Czarnecki, S. Ultrasonic Evaluation of the Pull-Off Adhesion between Added Repair Layer and a Concrete Substrate. *IOP Conf. Ser. Mater. Sci. Eng.* **2017**, *245*, 032037. [CrossRef]
61. Cai, G.; Zhao, J. Application of Sulphoaluminate Cement to Repair Deteriorated Concrete Members in Chloride Ion Rich Environment-A Basic Experimental Investigation of Durability Properties. *KSCE J. Civ. Eng.* **2016**, *20*, 2832–2841. [CrossRef]

materials

Article

The Use of NDT Diagnostic Methods and Calculations in Assessing the Masonry Tower Crowned with the Steel Dome

Krzysztof Grzyb *, Łukasz Drobiec, Julia Blazy and Jakub Zając

Department of Building Structures, Faculty of Civil Engineering, Silesian University of Technology, Akademicka 5, 44-100 Gliwice, Poland
* Correspondence: krzysztof.grzyb@polsl.pl

Abstract: Non-destructive testing (NDT) methods are a diagnostic tool for evaluating the risk of failure or the need for repair and renovation. In analyzing constructions of high historical value, destructive diagnostic methods should be avoided. This study is a comprehensive NDT investigation of the masonry tower topped with a steel dome, a remnant of the overhead telecommunications network from the end of the 19th century. Visual inspection and research made it possible to assess the degree of damage to the structure. Stress–strain state analysis showed the sufficient load-bearing capacity of the steel dome. In addition, calculations have shown that the masonry tower is subjected to significant horizontal forces causing structure cracks.

Keywords: non-destructive technique; visual inspection; ultrasonic testing; hardness testing; historic structures; cultural heritage; steel dome

Citation: Grzyb, K.; Drobiec, Ł.; Blazy, J.; Zając, J. The Use of NDT Diagnostic Methods and Calculations in Assessing the Masonry Tower Crowned with the Steel Dome. *Materials* **2022**, *15*, 7196. https://doi.org/10.3390/ma15207196

Academic Editors: Enrique Casarejos and Tetsuo Shoji

Received: 8 August 2022
Accepted: 10 October 2022
Published: 15 October 2022

Publisher's Note: MDPI stays neutral with regard to jurisdictional claims in published maps and institutional affiliations.

1. Introduction

One of the most significant challenges for sustainable development and cultural heritage protection is keeping historical buildings in the best possible condition [1–3]. Well-preserved cultural heritage is essential, not only from a historical point of view, but because it also decides the city's attractiveness and influences the life and safety of its habitants [4,5]. As a result, repair and renovation works are part of the life cycle of historic buildings to avoid deterioration, ageing processes, and eventual collapse [6,7]. The efficient maintenance of cultural heritage requires experience, broad knowledge, and a sense of aesthetics [8].

Non-destructive testing (NDT) methods are often used as a diagnostic tool for structures to inform about the risk of failure and the necessity of repair and renovation [9–11]. Their popularity in the examination of historic structures derives from the fact that NDT methods detect defects and evaluate the conditions of the materials without a necessity to destroy the studied element [12]. Additionally, NDT can be carried out not only during the exploitation of the object, but also during the construction stages to inform about its compliance with assumed requirements and properties [13].

This article presents the case study of the diagnosis and assessment of the historic post office building in Chorzów, Poland, which is around 130 years old. The methodology of the investigation process is divided into several stages, which is shown in Figure 1. The first step, while estimating the conditions and problems of the structure, is always connected with the visual assessment, which leads to the selection of the research methods. Later, the experimental investigation is provided to collect all the necessary information about the used materials and damage degree of the structure. The authors chose to use NDT methods because the studied building is a part of cultural heritage. Namely, hardness testing to check the grade of the structural steel was used because of its popularity and ease of implementation. Ultrasonic testing and microscopic methods were applied to the evolution of corrosion rate and thickness loss.

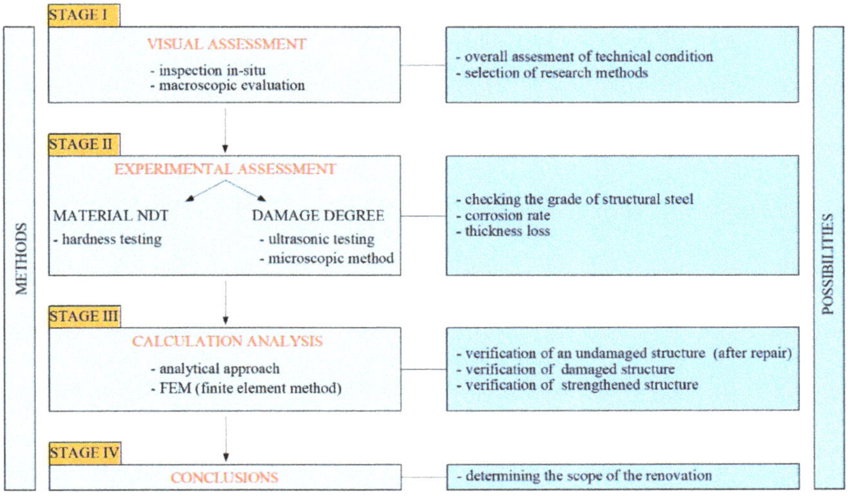

Figure 1. Stages of analysis of the technical condition of the structure.

1.1. Visual Testing

Visual testing is the oldest known NDT method and should be carried out as the first of all the NDT methods planned for diagnostics and testing [14,15]. It allows access to the following: the condition of the external surface of the element (surface method); the presence of any discontinuities (cracks, pores, voids, undercutting, sticking); shape defects; deformations; dimensional deviations; incorrect assembly; and operational damages (corrosion, fatigue erosion, leakage) [16,17]. To determine the element's dimensions or observed discontinuity, measuring devices such as calipers, depth gauges, and weld gauges are usually used [14]. Additionally, inspection mirrors, borescopes, endoscopes, periscopes, videoscopes, or drones may be required while diagnosing hard-to-reach places. Moreover, to observe the microstructure of the element microspores are used and when detailed analysis is needed, scanning electron microspores (SEM) are used [18–20]. It is also worth mentioning that the visual testing method is usually applied with some other NDT method.

1.2. Hardness Testing

Hardness testing distinguishes static and dynamic methods [21]. Within the former, the most popular ones are Brinell, Rockwell, and Vickers [21–24]. In all of them, the material hardness depends on force loading the indenter and permanent deformation caused by the action of this force. On the other hand, in the dynamic method, the hardness is measured based on the impact effect of the indenter on the researched surface. This method is relatively fast and easy to perform and, as a result, only minor damage to the specimen occurs, namely minor marks which are left on the tested surface [25]. In Shore's dynamic method, a steel weight ended with a hard indenter freely falling from a certain height, and the height of reflection from the tested material is measured [16]. It is noteworthy that some dynamic methods require a minimum mass and wall thickness of the tested object [26]. In [27], Useinov et al. studied the differences in hardness results at micro and nanometer scales. On the other hand, Huber and Heerens [28] highlighted the influence of a general residual stress state on indentation and hardness testing. Finally, the effect of steel heat treatment on hardness results was observed in [29].

1.3. Ultrasonic Testing

Ultrasonic testing (UT) is a volumetric method that yields the detection of discontinuities inside the tested object, its size, and its location [14,18]. Thanks to this tool, it is possible to receive information about the state of its interior and microstructure [22]. A bundle of ultrasonic waves is introduced into the element [14]. Then, due to the reflection from material discontinuities, it gives indications on the screen of the ultrasonic device. Afterward, the location and size of the detected discontinuities can be assessed based on the obtained information. It must be mentioned that the UT method applies to materials with good acoustic characteristics, allowing the ultrasonic wave to propagate [30]. These are materials with an isotropic and fine-grained structure. The most dangerous internal defects in materials and welds (cracks, tears, sticking, loss of penetration) can be detected using the UT method [16]. Moreover, it is possible to measure the material's thickness when it is accessible from only one side [16]. This is conducted by analyzing the time the wave travels through the studied element [31,32]. Additionally, modern thickness gauges yield accurate measurements without the need to remove the paint coating [33]. The ultrasonic leak testing also belongs to the UT method and gives information about the location and size of the leak. It is predominantly used to examine structures with high requirements regarding tightness, such as ships and tanks. In this method, the ultrasonic waves are emitted inside the object and if any discontinuity exists, the waves leave through them, recorded by the receiver outside the structure [34,35]. Nowadays, advanced UT techniques, such as UT phased array and UT TOFD, are more widely used. The former is based on introducing a bundle of ultrasonic waves excited in different configurations with phase shifts into the element, followed by their reception, while preserving the previous phase shifts [14]. This method is often used for elements with complex shapes and weld joints, as well as to monitor corrosion processes and create corrosion maps [36]. On the other hand, in the UT TOFD method, a bundle of ultrasonic waves is introduced to excite diffraction waves on the existing material discontinuities. Then, it is received and different flight times are measured [14]. It main application is found in examining the presence of cracks and their size in welded joints [37]. Furthermore, it is characterized by high detectability, a limited number of false readings, accurate geometry measurements of discontinuity, and a short execution time [38]. In the literature, Lin et al. proposed to use advanced UT techniques to evaluate the steel-bridge weld joints [39]. Additionally, due to its high testing speed, high efficiency, and immediately obtained results after the test, UT is one of the most commonly used NDT methods in the evaluation of steel structures, during both manufacturing and exploitation [40,41]. However, as every method of UT has its limitations and for increased roughness of the external surface of the object, the testing possibilities decrease.

A summary of the selected NDT method's advantages and limitations, together with application indications, is shown in Table 1 [42,43].

The next stage in assessing the historic structure's technical condition after a non-destructive diagnosis is connected to calculations and numerical analyses. There are three situations in which the exploited building may be verified. Namely, the analysis of an undamaged, damaged, or strengthened structure can be conducted (Figure 1). It must be mentioned that the steel structure has to resist forces from self-weight, live load, dead load, wind load, snow load, and accidental load. Besides this, changes in the structure's temperature resulting from summer heating and winter cooling generate significant material stresses [44]. As a result, they must be considered while analyzing the element. Lien et al. [45] studied the non-linear behavior of steel structures subjected to heating and cooling stages. They confirmed that the elevated temperatures greatly influence the deformations of the steel frame, including deflections and displacements. Moreover, the temperature of the steel structure varies from the surrounding air temperature and can be non-uniformly distributed within the element [46,47]. Another important issue concerns the temperature at which the object was constructed because this phenomenon directly influences the thermal stresses induced in the material during exploitation [48]. Regarding the material properties, it should be considered that steel tensile and yield strength decreases with increasing temperatures [49,50]. However, it must be highlighted that both high and low temperatures have a negative influence on steel structures. Namely, when cooling down, steel becomes more brittle and losses its ductility [51,52], which is dangerous for the structure and may lead to failure. In conclusion, the problem of extreme temperatures has to be appropriately considered and addressed while analyzing steel structures.

Table 1. Advantages and limitations of selected non-destructive testing methods.

NDT Methods	Advantages	Limitations
Visual testing	• Portable • Inexpensive • Commonly known • Easy to implement • Minimum preparation time • Little or no special equipment	• Only large flaws possible to detect • Misinterpretations when disruptive optical effects or poor visuality
Hardness testing	• Portable • Commonly known • Easy to implement • Can be adapted to various types of materials	• Requirements for surface preparation • Possible deformations due to cutting and grinding • Influenced by various factors
Ultrasonic testing	• Portable • Low costs • Rapid results • High sensitivity and detectability • Flaw detection deep in the element • Some capacity to access the location, size, nature, and shape of the discontinuity • Applicable also for elements with complicated shape	• Long detection time • Need for coupling agent • Difficult to study small, thin, and complex elements • Good acoustic characteristics of tested material required • For relatively smooth surfaces, limitations when roughness increases

Finally, a proper assessment of the technical condition of the structure leads to a statement about the further possibility of the functioning of the structure (Figure 1). The hazard of failure, or even collapse, should be eliminated not only for the safety of people, but also for surrounding buildings. Additionally, when it comes to the historic structure, the cultural heritage value should be taken into account. Therefore, the comprehensive assessment of the structure should be a combination of tests, measurements, engineering calculations, numerical analysis, and finally, guidelines for repair works. All of these stages were applied in assessing the technical conditions of a steel dome built on top of a brick tower of a post office in Chorzów, Poland.

The contribution of the presented case study research relies on showing the practical implementation of the theoretical background. The paper indicates the successive steps of diagnosis and evaluation of the historical building that should be taken. The analysis reveals the importance of a holistic approach during structural analysis. The article presents how experimental and numerical analysis yields the consideration of phenomena that were neglected in the past. Namely, the temperature influence was not considered while designing the steel dome built on top of the brick tower. The research significance of the work also appears in presenting specific renovation solutions for the structure. Nowadays, more and more structures designed for a 50- or 100-year service life require an assessment of the technical conditions. The article suggests the research methodology and presents a comprehensive case study.

2. Case Study

2.1. Historical Background

The building was erected in 1891–1892 in the neo-gothic style, according to the design of J. Schubert. The building was expanded in 1911 according to the design of F. Nhagen (a wing called the parcel department was added). The post office was entered into the register of monuments on 30 October 1984. Since its construction, the building has been one of the city's landmarks.

2.2. Architectural Values and Building Structure

The building is a two-story building, topped with a hipped roof with dormers. The building was erected on an irregular plan. The brick tower was built on an octagonal plan, ending with the subject steel dome with a spire. The element crowning the tower is a remnant of the overhead telecommunications system from the end of the 19th century. Apart from the architectural role, the steel elements of the dome also played the role of a complex antenna system in itself, which was unique at the time of its construction. The facade is made of red clinker bricks, decorated with glazed green and brown bricks. Figure 2a shows the entrance to the post office. The binding of masonry elements with colored masonry elements, which are decorated around the window openings, is shown in Figure 2b.

(**a**) (**b**)

Figure 2. Architectonic details. (**a**) The facade of the building with a visible main entrance; (**b**) decorations in the form of colored bricks around the window opening.

The brick part of the tower is topped with a flat wooden roof. A steel hatch in the flat roof leads to the interior of the openwork structure of the dome. The construction of the cupola consists of different rod elements. The main load-bearing elements include eight steel ribs (Figure 3). The rib cross-section consists of two arches, stiffened at the ends with flat bars. The maximum width of the cross-section is 38 cm (Figure 4a), and the thickness of the rib is, on average, 15 mm. In the place where the ribs are constrained, U-shaped stiffening sheets are inbuilt. In the lower part of the dome, the ribs are connected by a decorated steel balustrade (Figure 4b). Above each intercostal segment, nine L45 × 45 angles were made. The top of the dome consists of steel pipes and flat bars that support the bell, crowning the helmet as a whole.

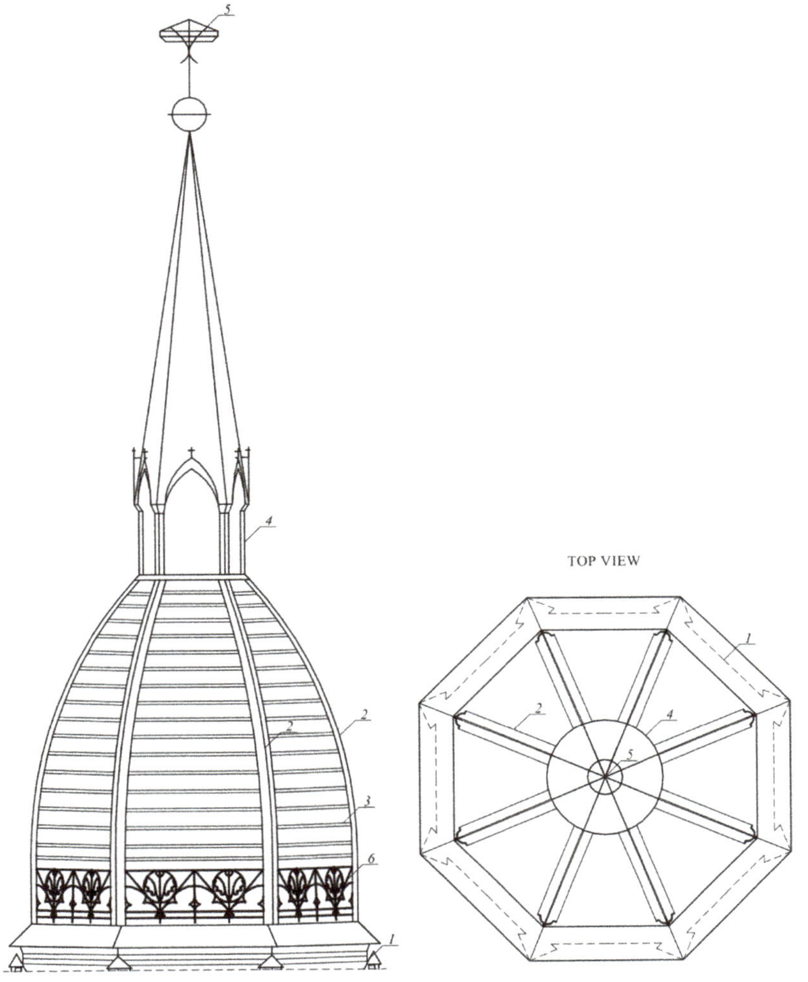

Figure 3. Side and top view of the dome: *1*—brick tower; *2*—ribs; *3*—angles; *4*—helmet; *5*—spire; *6*—decorative railing.

(**a**) (**b**)

Figure 4. Steel dome structure. (**a**) View of the balustrade surrounding the dome; (**b**) the 38 cm width of one of the dome's eight supporting ribs.

3. The Technical Condition and Diagnostic Methods

3.1. Macroscopically Observed Abnormalities

During on-site inspections carried out at the facility, certain damages may occur, including the corrosion of dome structure elements, incorrect connections of the dome structure elements, local constrictions of load-bearing elements, corrosion of steel built around the flat wooden roof, cracks on the tower's masonry structure, and overgrowing vegetation on the tower body (on the facade).

Paint chips were visible on the dome's structural elements, partly due to the corrosive processes under the paint coating (Figure 5a). The unprotected steel parts of the helmet were exposed to unfavorable weather conditions, including rainwater or moisture condensing on the steel surface. The apparent lack of proper structure protection with paint coatings led to the intensification of corrosion processes and a reduction in the effective cross-section of the element involved in the transfer of loads.

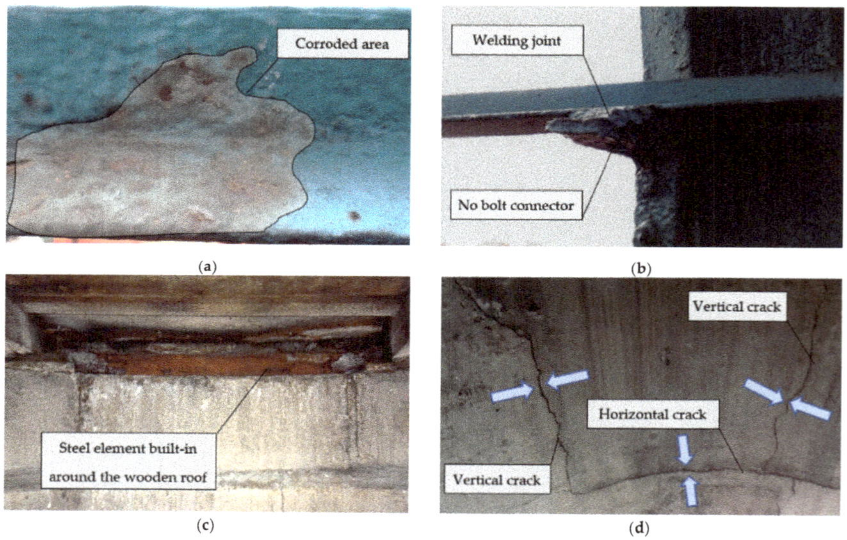

Figure 5. Technical problems negatively impact the structure's technical condition. (**a**) Corrosion of the L45 × 45 angle, and visible bursting of the painted surface with corrosion products; (**b**) support for the L45 × 45 angle section at the connection with the rib; (**c**) corrosion of steel profile around the flat wooden roof; (**d**) cracks in the masonry structure below the flat wooden roof, and vertical and horizontal damages; *the arrows indicate cracks.

The structure inspection showed that the dome elements' connections were mainly designed to be bolted (or riveted). In the present state, all connection points are covered with paint, and locally in some places, it is visible that the elements were welded together at the point of connections with the use of connectors (bolts/rivets). This was supposed to "strengthen" or "preserve" the connection, but such a solution cannot be considered correct (Figure 5b). Moreover, undesirable constrictions of the dome's structural elements were locally visible, which reduces the load-bearing capacity of the bars in these places.

The roof, which separates the part of the brick tower from the steel, openwork helmet of the dome, is a wooden-beam structure. The technical condition of the flat roof can be considered reasonable; however, the dampness in the place where the rest of the beam is located is disturbing. A steel element runs around the flat roof. It was found that this highly corroded profile may suggest the accumulation of moisture in this place. This element is probably the lower circumferential brace of the dome. Figure 5c shows the high degree of corrosion of this element.

Below the flat wooden roof, numerous cracks in the masonry structure are visible from the inside of the tower. The damages are visible to the naked eye from the internal steel staircase. The cracks are vertical and horizontal (Figure 5d) and run along the circumference of the tower body. In the corners of the openings, diagonal cracks with a smaller width are visible—these are places of stress concentration resulting from lower stiffness. The damages are located below the flat wooden roof up to the upper edge of the window openings (a few meters worth of crack strips). The crack opening is at least 2.0 mm.

The abnormalities also include vegetation on the facade of the tower body, which can lead to the biological corrosion of the masonry structure or significant wall dampness in these places.

3.2. *Non-Invasive Structure Diagnostics and Calculations*

Various non-destructive diagnostic methods were used to determine the damage to the structure and establish the causes of abnormalities. Microscopic imaging enabled the assessment of the degree of corrosion of the steel elements. The analyzed areas were photographed with 50-times zoom. Such an enlargement made it possible to locate micro-damage and cracks in the steel.

The actual thickness of the elements was verified using the ultrasonic method. The tests were carried out using the SONO M460 device by Metrison, Poland. The device is an ultrasonic thickness gauge designed to perform very precise, fast, and non-destructive measurements in the range of 0.65 mm to 600 mm. The device works with many heads that measure the thickness of the walls of pipes, tanks, and structural elements, both in the range of small thicknesses and those whose temperature can reach up to 300 °C, as well as corroded objects and those covered with a protective layer.

The device allows for the smooth adjustment of the ultrasonic wave speed according to the tested materials in a range from 1000 to 9999 m/s. It is also possible to use ready-programmed wave-speed settings for 22 types of materials, including the basic ones, such as steel, non-magnetic steel, aluminum, brass, copper, iron, cast iron, gold, silver, zinc, tin, nickel, titanium, epoxy resin, plexiglass, P.V.C., and ceramics. To obtain an even more accurate measurement result, the SONO M460 (Metrison, Poland) needs calibration performed by the user on samples of the measured materials. The thickness gauge used in the tests was equipped with a MEE5 5MHz ultrasonic head, which enables measurement in the range 2.5–100 mm.

The hardness tests were used to determine the actual steel grade of the cupola structure. Sauter HMM (Balingen, Germany), a type D device, was used. A rebound sensor was driven against the surface of the tested element with spring force. Depending on object hardness, the kinetic energy of the module is absorbed, and the speed reduction is measured and converted to the hardness values (Figure 6).

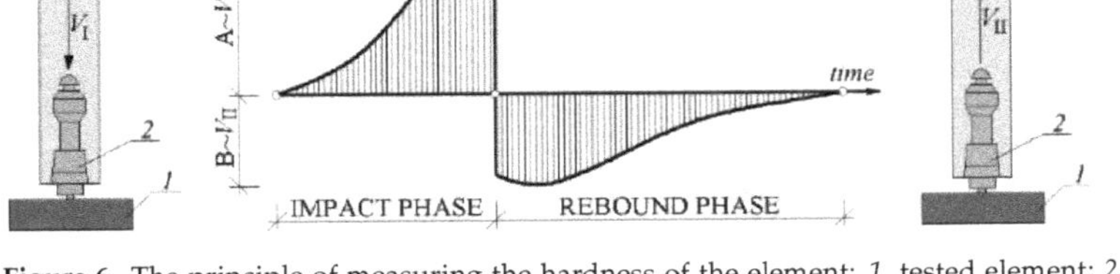

Figure 6. The principle of measuring the hardness of the element: *1*, tested element; *2*, movable mandrel; *3*, housing of the device; *4*, interior spring; V_I, velocity in impact phase; V_{II}, velocity in the rebound phase.

The voltages were proportional to the velocities and were processed and displayed as the hardness value 'L' (Leeb value). The surface of the tested steel should be flat with a metallic luster, and with no oxide skin or other dirt. In practice, obtaining such a surface is not easy, especially if there are corrosion losses in the steel (Figure 7). It should be added that the test sample should also be hard and stiff; otherwise, it may cause erroneous results.

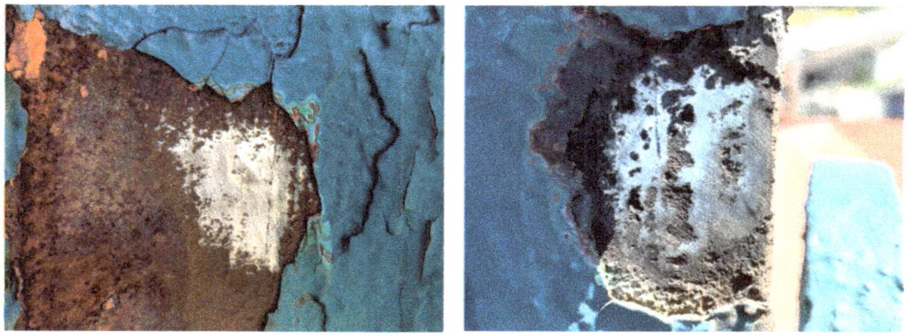

(**a**) (**b**)

Figure 7. Cleaning the steel structure of paint coatings for hardness testing. (**a**) Tested area 1 of the main rib; (**b**) tested area 2 of the main rib.

Additionally, static calculations were performed on the spatial model to verify the stress state in the dome structure. The structural steel grade S235 was assumed. The structure was subjected to self-weight, wind, and climatic temperature loads. The value of the initial temperature T_0 in Poland was assumed as 8 °C. The maximum heating temperature predicted in the calculations was 36 °C, and the cooling temperature was 32 °C. The calculations were also corrected, taking into account the height of the building, located at a height of 264 m above sea level. The analysis was made following the recommendations of the Eurocode. The cross-sections of the main load-bearing elements and the structure's geometry were prepared based on an inventory made during on-site inspections (Figure 8). The loads were applied using load-panel-surface elements that transfer loads to structural elements. Calculations make it possible to assess the stress–strain state of a structure, thus determining the scope of the strengthening of structural elements.

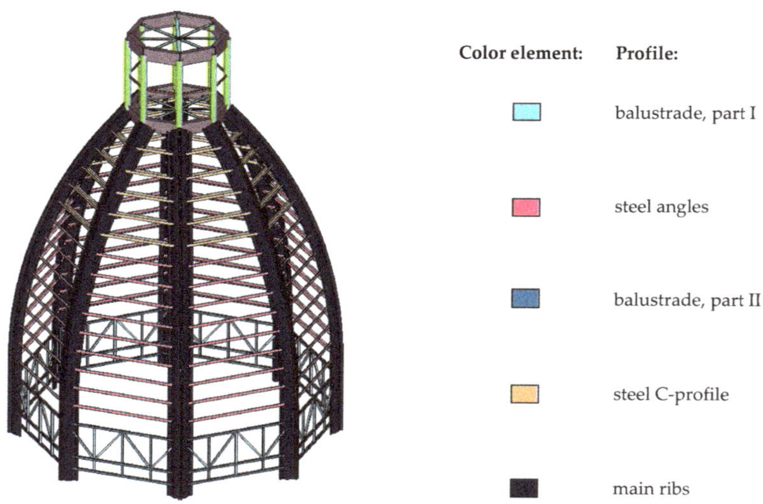

Figure 8. View of the spatial computational model.

4. Results

4.1. Results of Non-Destructive Testing

The hardness tests of structural steel in four selected places were analyzed. The result of the test was single rebound values (Leeb hardness number). The measurement results are presented in Figure 9.

Each measurement point was tested with eight approaches. The apparatus required at least five correct rebound results to determine the tensile strength and yield point correctly. If the measurement result was unreliable (gray color in Figure 9), it was not included in the estimation of mechanical parameters. The maximum standard deviation for the six rebounds was obtained at the second measuring point and amounted to 8.64. The yield point was obtained in two test points (first and second). The grade of the built-up steel was S235, although the results that were obtained suggest that the steel's yield strength was higher and lower. Analyzing steels with alloys different from those currently used required expert assessment, particularly when applying the currently applicable standards, including assigning steels to current strength categories. Microscopic imaging made it possible to estimate the size of and damage to the structure of the steel structure. Thanks to this, it was possible to exclude cracks in the steel itself, and the cracks were mainly related to the paint coating. Moreover, remnants of another paint were discovered as shown in Figure 10.

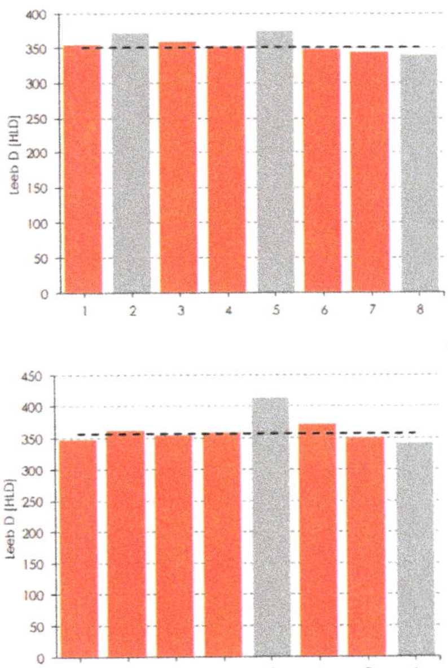

Statistics:
-number of measurements: n = 5,
-mean value of the measurement: Lmv = 352 HLD,
-standard deviation: s_L = 6.20.
Strength parameters of the material:
-steel according to PN-EN 10210-1: S235,
-tensile strength $f_u(R_m)$ = 360 MPa,
-yield point $f_y(R_{eH})$ = 235 MPa.

(**a**)

Statistics:
-number of measurements: n = 6,
-mean value of the measurement: Lmv = 357 HLD,
-standard deviation: s_L = 8.64.
Strength parameters of the material:
-steel according to PN-EN 10210-1: S235,
- ensile strength $f_u(R_m)$ = 360 MPa,
-yield point $f_y(R_{eH})$ = 235 MPa.

(**b**)

Figure 9. *Cont.*

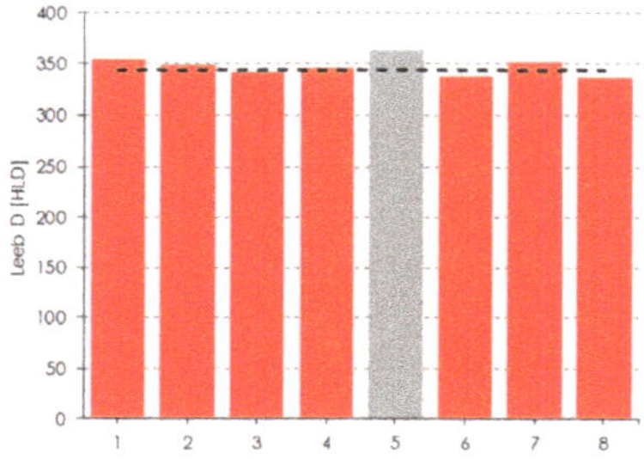

Statistics:

-number of measurements: $n = 7$,

-mean value of the measurement: Lmv = 345 HLD,

-standard deviation: $s_L = 6.74$.

Strength parameters of the material:

-steel outside the grade,

-tensile strength $f_u(R_m) = 319$ MPa,

-yield point $f_y(R_{eH}) = 207$ MPa.

(**c**)

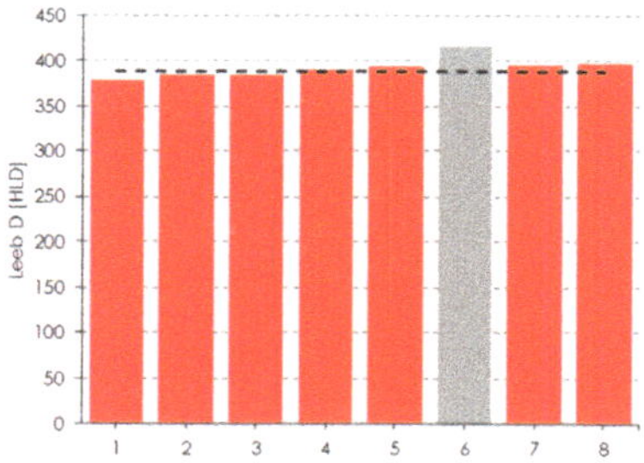

Statistics:

-number of measurements: $n = 7$,

-mean value of the measurement: Lmv = 389 HLD,

-standard deviation: $s_L = 6.96$.

Strength parameters of the material:

-steel according to PN-EN 10210-1: S275,

-tensile strength $f_u(R_m) = 430$ MPa,

-yield point $f_y(R_{eH}) = 275$ MPa.

(**d**)

Figure 9. Test results for the steel structure of the dome. (**a**) Measuring point 1; (**b**) measuring point 2; (**c**) measuring point 3; (**d**) measuring point 4; gray color, unreliable rebound result omitted in the estimation of mechanical parameters.

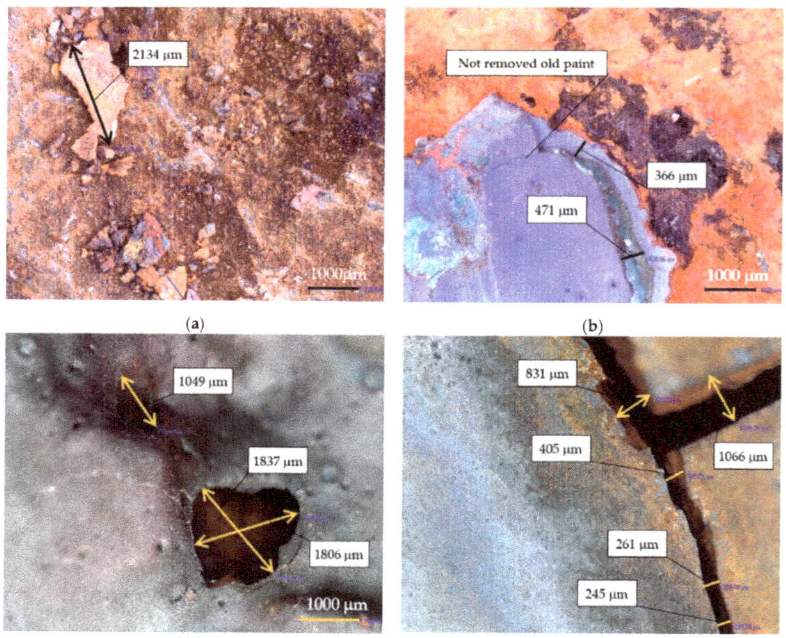

Figure 10. Microscopic imaging results. (**a**) Measured the size of the corrosion product 2134 μm; (**b**) thickness of the corrosion pitting 366 μm and 471 μm; (**c**) damage to paint coating, 1806 × 1837 μm size; (**d**) cracks in the paint coating, maximum size 1066 μm.

Based on ultrasonic tests, the size of corrosion damage was estimated. In some cases, the cross-section loss reached up to 11% (Table 2).

Table 2. Results of ultrasonic tests-measurement of the actual thickness of steel angles.

Tested Steel Angle	Element 1	Element 2	Element 3
Measured thickness range	5.98–6.00 mm	4.75–5.38 mm	5.65–5.67 mm
Average thickness	5.99 mm	5.31 mm	5.66 mm
Corrosive loss	1%	11%	6%

4.2. Results of Calculations

The stress method was adopted, comparing the maximum stresses with the assumed structural steel yield strength. The magnitudes of stresses in the main load-bearing element ribs are shown in Figure 11 and Table 3. Stresses for the other steel elements are shown in Figure 12.

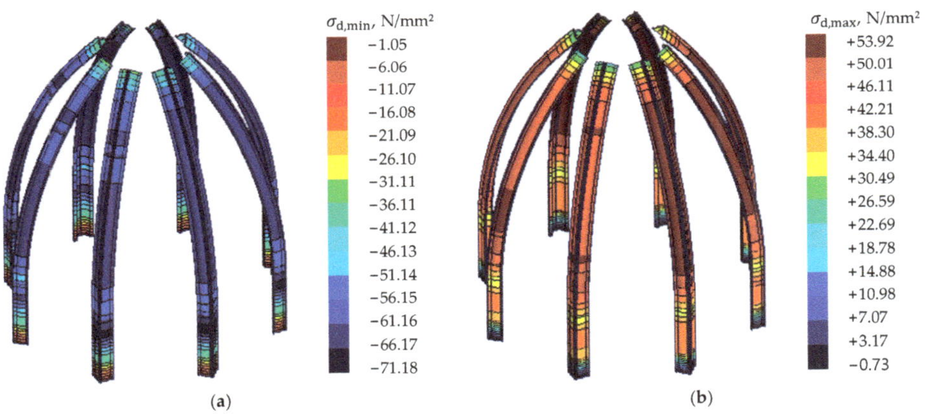

Figure 11. Stresses for main ribs of cupola. (**a**) minimum values; (**b**) maximum values.

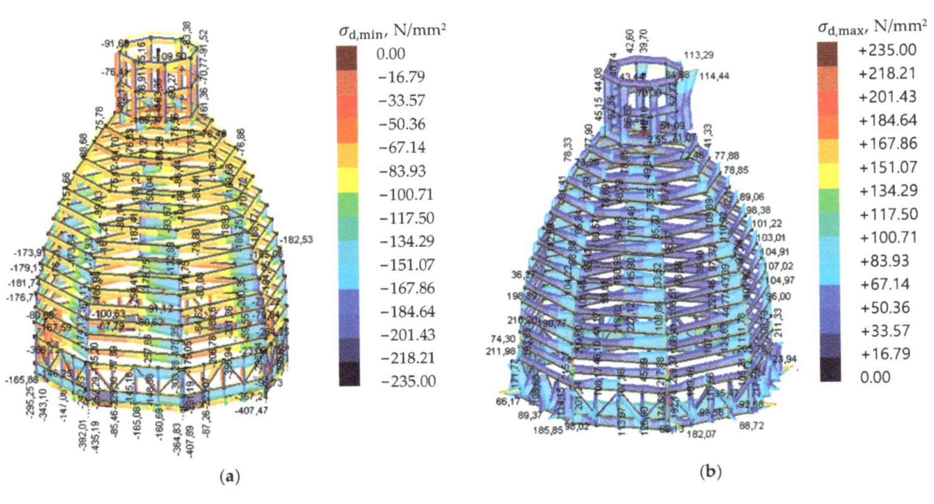

Figure 12. Stresses for the other steel elements. (**a**) minimum values; (**b**) maximum values.

Table 3. A summary of the tensile and compressive stresses of the main ribs of the dome.

Maximum Stress, $\sigma_{d,max}$	Minimum Stress, $\sigma_{d,min}$	Stress-Strain State, $\lvert\sigma_d\rvert/f_{yd}$
53.92 N/mm^2	-	0.23
-	−71.18 N/mm^2	0.30

It can be assumed that the design stresses in the ultimate limit state (ULS) along the length of the bars did not exceed the steel yield stress (local exceedances exist at the point of connection of the elements with the steel ribs). Stress concentration occurs in the connection between particular elements. The structural arrangement of the dome insures the main ribs against buckling. The structure is not susceptible to a global or local loss of stability. The structure's temperature changes, resulting from heating the elements in summer or cooling them down in winter, generating significant forces in the dome-support area. Reaction value forces in the supporting area are shown in Figure 13.

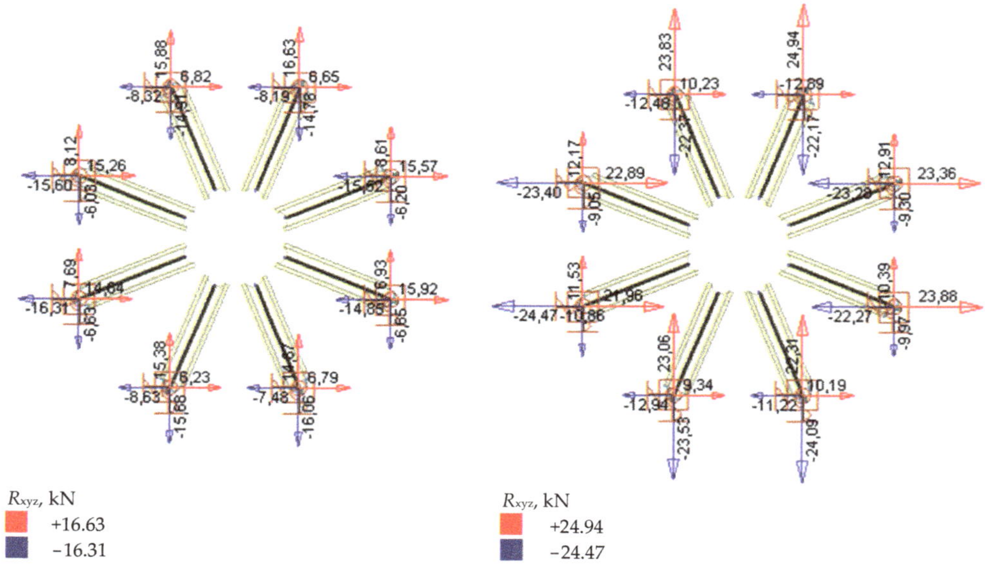

(**a**) (**b**)

Figure 13. Reactions in the supporting area. (**a**) characteristic values (**b**) design values.

5. Discussion and Conclusions

The study demonstrates the usefulness of non-destructive methods in assessing a structure of unique historical value. The preliminary stage for the evaluation of the scope of the research was an on-site inspection. Then, structural steel hardness tests were carried out to assign the steel grade. The calculations made it possible to verify the load-bearing capacity of the elements.

During the local inspections, numerous cracks in the masonry tower were found. The morphology of the cracks indicated the initiation of tensile stresses on the inside of the tower. The vertical circumferential cracks suggested that forces were transmitted where the steel dome is attached. Nevertheless, damage to the masonry led to the degradation of the monument and lowered the stiffness of the entire tower structure.

These problems can result from the thermal stresses generated in the structural elements of the steel dome during intense heating or cooling. The external factors are determined by season. High stresses were, in the original form, to be taken by the circumferential steel ring connected with the main ribs of the dome. However, the existing perimeter profile in the flat roof level is highly corroded and locally cracked. Its structural function is not fulfilled. The perimeter beam's crack should be considered a direct cause of visible cracks in the masonry structure.

Therefore, making a new ring that supports and absorbs the horizontal forces is necessary before starting repairs to the masonry structure. The tower's masonry structure should be preserved to limit the propagation of cracks or new damage. The conducted analysis shows how important a holistic approach to structural diagnosis is. During the research, it may turn out that the observed abnormalities result from many different factors. Identifying the causes of damage to historic structures is a challenging and not obvious task. Non-destructive diagnostic methods fit in with the approach that is a compromise between preserving the complete historical form in every aspect and the need to verify the structure.

The repair of the building will consist of recreating the steel-ring top beam below the dome, which prevents the transfer of horizontal forces to the masonry structure. An additional reinforcement in the bed joints of the masonry tower structure would ensure the crack opening's limitation.

Author Contributions: Conceptualization, K.G. and Ł.D.; methodology, J.B.; software, K.G. and J.Z.; validation, J.Z. and Ł.D.; formal analysis, K.G. and J.Z.; investigation, J.Z. and Ł.D.; resources, J.B.; data curation, J.B.; writing—original draft preparation, K.G. and J.B.; writing—review and editing, J.Z.; visualization, K.G. and J.B.; supervision, Ł.D. All authors have read and agreed to the published version of the manuscript.

Funding: This publication was supported by the own resources of the Department of Building Structures at the Silesian University of Technology.

Institutional Review Board Statement: Not applicable.

Data Availability Statement: Not applicable.

Acknowledgments: Krzysztof Grzyb is a holder of a European Unionscholarship through the European Social Fund, grant InterPOWER (POWR.03.05.00-00-Z305).

Conflicts of Interest: The authors declare no conflict of interest.

References

1. Ruggiero, G.; Marmo, R.; Nicolella, M. A Methodological Approach for Assessing the Safety of Historic Buildings' Façades. *Sustainability* **2021**, *13*, 2812. [CrossRef]
2. Rodrigues, F.; Matos, R.; Di, M.; Costa, A. Conservation level of residential buildings: Methodology evolution. *Constr. Build. Mater.* **2018**, *172*, 781–786. [CrossRef]
3. Huynh, T.; Nguyen, B.-P.; Pradhan, A.M.S.; Pham, Q.-Q. Vision-based inspection of bolted joints: Field evaluation on a historical truss bridge in Viet Nam. *Vietnam. J. Mech.* **2020**, *36*, 77. [CrossRef]
4. Borri, A.; Corradi, M. Architectural Heritage: A Discussion on Conservation and Safety. *Heritage* **2019**, *2*, 631–647. [CrossRef]
5. Du, F.; Okazaki, K.; Ochiai, C.; Kobayashi, H. Post-disaster building repair and retrofit in a disaster-prone historical village in China: A case study in Shangli, Sichuan. *Int. J. Disaster Risk Reduct.* **2016**, *16*, 142–157. [CrossRef]
6. Nowogónska, B. Consequences of Abandoning Renovation: Case Study—Neglected Industrial Heritage Building. *Sustainability* **2020**, *12*, 6441. [CrossRef]
7. Drobiec, Ł.; Grzyb, K.; Zając, J. Analysis of reasons for the structural collapse of historic buildings. *Sustainability* **2021**, *13*, 10058. [CrossRef]
8. Cardani, G.; Belluco, P. Reducing the Loss of Built Heritage in Seismic Areas. *Buildings* **2018**, *8*, 19. [CrossRef]
9. Yi, J.; Park, W.: Lee, S.; Huynh, T.; Kim, J.; Seo, C. Evaluation of Vibration Characteristics of an Existing Harbor Caisson Structure Using Tugboat Impact Tests and Modal Analysis. *Int. J. Ofdistributed Sens. Netw.* **2013**, *2013*, 1–11. [CrossRef]
10. Balayssac, J.-P.; Garnier, V. (Eds.) *Non-Destructive Testing and Evaluation of Civil Engineering Structures*; Elsevier Science: Amsterdam, The Netherlands, 2017; ISBN 9780081023051.
11. Ramesh, G.; Srinath, D.; Ramya, D.; Vamshi Krishna, B. Repair, rehabilitation and retrofitting of reinforced concrete structures by using non-destructive testing methods. *Mater. Today Proc.* **2021**, in press. [CrossRef]
12. Reddy, K.A. Non-Destructive Testing, Evaluation Of Stainless Steel Materials. *Mater. Today Proc.* **2017**, *4*, 7302–7312. [CrossRef]
13. Deepak, J.R.; Raja, V.K.B.; Srikanth, D.; Surendran, H.; Nickolas, M.M. Non-destructive testing (NDT) techniques for low carbon steel welded joints: A review and experimental study. *Mater. Today Proc.* **2021**, *44*, 3732–3737. [CrossRef]

14. TÜV Rheinland. Non-Destructive Testing Methods. Available online: https://www.tuv.com/poland/pl/lp/tuv-rheinland-academy/main-navigation/training/szkolenia-w-polsce/badania-nieniszczące/?wt_mc=SEA.Ads.Google.PL22_A02_NDT.PL22_A02_NDT_GA.textad.google_ads_ndt&cpid=PL22_A02_NDT_GA (accessed on 21 June 2022).
15. Ghosh, N.; Kumar, P.; Nandi, G. Parametric Optimization of MIG Welding on 316L Austenitic Stainless Steel by Grey-Based Taguchi Method. *Procedia Technol.* **2016**, *25*, 1038–1048. [CrossRef]
16. Diagnostyka Techniczna-Badania. Non-Destructive Testing of Steel Materials and Their Connections. Available online: http://www.dt-b.pl/badania-nieniszczace-materialow-stalowych-i-ich-polaczen (accessed on 21 June 2022).
17. DRACO Laboratorium Badań Nieniszczących. Visual Method. Available online: https://www.draco.com.pl/en/our-offer/visual-method-vt (accessed on 5 July 2022).
18. Chen, B.; Wang, C.; Wang, P.; Zheng, S.; Sun, W. Research on Fatigue Damage in High-Strength Steel (FV520B) Using Nonlinear Ultrasonic Testing. *Shock. Vib.* **2020**, *2020*, 8847704. [CrossRef]
19. Xu, X.; Mi, G.; Xiong, L.; Jiang, P.; Shao, X.; Wang, C. Morphologies, microstructures and properties of TiC particle reinforced Inconel 625 coatings obtained by laser cladding with wire. *J. Alloy. Compd.* **2018**, *740*, 16–27. [CrossRef]
20. Chen, L.; Yuan, F.P.; Jiang, P.; Wu, X.L. Mechanical properties and nanostructures in a duplex stainless steel subjected to equal channel angular pressing. *Mater. Sci. Eng. A* **2012**, *551*, 154–159. [CrossRef]
21. Walley, S.M. Historical origins of indentation hardness testing. *Mater. Sci. Technol.* **2012**, *28*, 1028–1044. [CrossRef]
22. Kennedy, J.R.; Wiskel, J.B.; Ivey, D.G.; Henein, H. L80 pipe steel microstructure assessment using ultrasonic testing. *Mater. Sci. Technol.* **2019**, *35*, 1942–1949. [CrossRef]
23. Fydrych, D.; Łabanowski, J.; Rogalski, G.; Haras, J.; Tomków, J.; Świerczyńska, A.; Jakóbczak, P.; Kostro, Ł. Weldability of S500MC Steel in Underwater Conditions. *Adv. Mater. Sci.* **2014**, *14*, 37–45. [CrossRef]
24. Chandler, H. *Hardness Testing*; ASM International: Almere, The Netherlands, 1999.
25. emcoTEST. Principles of Hardness Testing. *Hardness Testing—A Material Testing Method.* Available online: https://www.emcotest.com/en/the-world-of-hardness-testing/hardness-know-how/theory-of-hardness-testing/principles-of-hardness-testing-54/hardness-testing-a-material-testing-method-230/ (accessed on 16 September 2022).
26. Navitest. Hardness Testing. Available online: https://navitest.com.pl/pl/badania-nieniszczace/badania-twardosci/ (accessed on 2 July 2022).
27. Useinov, A.; Gogolinskiy, K.; Reshetov, V. Mutual consistency of hardness testing at micro- and nanometer scales. *Int. J. Mater. Res.* **2009**, *100*, 968–972. [CrossRef]
28. Huber, N.; Heerens, J. On the effect of a general residual stress state on indentation and hardness testing. *Acta Mater.* **2008**, *56*, 6205–6213. [CrossRef]
29. Srivastava, K.; Sinha, A.A.; Sahani, R. Effect of heat treatment on hardness and toughness of EN8 steel. *Mater. Today Proc.* **2020**, *42*, 1297–1303. [CrossRef]
30. Nakahata, K.; Hirose, S.; Schubert, F.; Köhler, B. Image Based EFIT Simulation for Nondestructive Ultrasonic Testing of Austenitic Steel. *J. Solid Mech. Mater. Eng.* **2009**, *3*, 1256–1262. [CrossRef]
31. Navitest. Ultrasonic Thickness Measurements. Available online: https://navitest.com.pl/pl/badania-nieniszczace/ultradzwiekowe-pomiary-grubosci-uttutm/ (accessed on 28 June 2022).
32. Texas Nondestructive Testing Academy. Ultrasonic Testing. *Advantages & Disadvantages.* Available online: https://www.txndt.com/safety-section/ultrasonic-testing (accessed on 16 September 2022).
33. DRACO Laboratorium Badań Nieniszczących. Ultrasonic Thickness Testing. Available online: https://www.draco.com.pl/en/our-offer/ultrasonic-thickness-testing-utt (accessed on 5 July 2022).
34. DRACO Laboratorium Badań Nieniszczących. Ultrasonic Leak Testing. Available online: https://www.draco.com.pl/en/our-offer/ultrasonic-leak-testing-lt (accessed on 5 July 2022).
35. Navitest. Ultrasonic Leak Testing. Available online: https://navitest.com.pl/pl/badania-nieniszczace/ultradzwiekowe-pomiary-szczelnosci/ (accessed on 5 July 2022).
36. DRACO Laboratorium Badań Nieniszczących. Ultrasonic Tests Phased Array. Available online: https://www.draco.com.pl/en/our-offer/ultrasonic-tests-phased-array-pa (accessed on 5 July 2022).
37. Chassignole, B.; Duwig, V.; Ploix, M.A.; Guy, P.; El Guerjouma, R. Modelling the attenuation in the ATHENA finite elements code for the ultrasonic testing of austenitic stainless steel welds. *Ultrasonics* **2009**, *49*, 653–658. [CrossRef] [PubMed]
38. DRACO Laboratorium Badań Nieniszczących. Time of Flight Diffraction. Available online: https://www.draco.com.pl/en/our-offer/time-of-flight-diffraction-tofd (accessed on 5 July 2022).
39. Lin, Z.B.; Azarmi, F.; Al-Kaseasbeh, Q.; Azimi, M.; Yan, F. Advanced Ultrasonic Testing Technologies with Applications to Evaluation of Steel Bridge Welding—An Overview. *Appl. Mech. Mater.* **2015**, *727–728*, 785–789. [CrossRef]
40. Navitest. Ultrasonic Testing. Available online: https://navitest.com.pl/pl/badania-nieniszczace/badania-ultradzwiekowe/ (accessed on 28 June 2022).
41. Dombret, P. Methodology for the ultrasonic testing of austenitic stainless steel. *Nucl. Eng. Des.* **1991**, *131*, 279–284. [CrossRef]
42. Thorat, S.; Non Destructive Testing. NDT Testing. *Procedure Of NDT.* Available online: https://learnmech.com/non-destructive-testing-ndt-testing-procedure-of-ndt/ (accessed on 16 September 2022).
43. Dwivedi, S.K.; Vishwakarma, M.; Soni, P.A. Advances and Researches on Non Destructive Testing: A Review. *Mater. Today Proc.* **2018**, *5*, 3690–3698. [CrossRef]

44. Liu, H.; Chen, Z.; Zhou, T. Investigation on temperature distribution and thermal behavior of large span steel structures considering solar radiation. *Adv. Steel Constr.* **2013**, *9*, 41–58.
45. Lien, K.H.; Chiou, Y.J.; Wang, R.Z.; Hsiao, P.A. Nonlinear behavior of steel structures considering the cooling phase of a fire. *J. Constr. Steel Res.* **2009**, *65*, 1776–1786. [CrossRef]
46. Liu, H.; Chen, Z.; Zhou, T. Theoretical and experimental study on the temperature distribution of H-shaped steel members under solar radiation. *Appl. Therm. Eng.* **2012**, *37*, 329–335. [CrossRef]
47. Becker, R. Structural behavior of simple steel structures with non-uniform longitudinal temperature distributions under fire conditions. *Fire Saf. J.* **2002**, *37*, 495–515. [CrossRef]
48. Liu, H.; Chen, Z.; Chen, B.; Xiao, X.; Wang, X. Studies on the temperature distribution of steel plates with different paints under solar radiation. *Appl. Therm. Eng.* **2014**, *71*, 342–354. [CrossRef]
49. Sun, R.; Huang, Z.; Burgess, I.W. Progressive collapse analysis of steel structures under fire conditions. *Eng. Struct.* **2012**, *34*, 400–413. [CrossRef]
50. Outinen, J.; Makelainen, P. Mechanical properties of structural steel at elevated temperatures and after cooling down. *Fire Mater.* **2004**, *28*, 237–251. [CrossRef]
51. Leroy, C.; Alvaro, A.; Maljaars, J. The effect of low temperatures on the fatigue crack growth of S460 structural steel. *Int. J. Fatigue* **2016**, *82*, 110–118. [CrossRef]
52. Paik, J.K.; Kim, B.J.; Park, D.K.; Jang, B.S. On quasi-static crushing of thin-walled steel structures in cold temperature: Experimental and numerical studies. *Int. J. Impact Eng.* **2011**, *38*, 13–28. [CrossRef]

materials

Article

The Influence of Biofilm on Selected Properties of Thin-Coat Mineral-Based Plasters on EPS Substrate

Monika Dybowska-Józefiak and Maria Wesołowska *

Faculty of Civil and Environmental Engineering and Architecture, Bydgoszcz University of Science and Technology, 85-796 Bydgoszcz, Poland
* Correspondence: maria.wesolowska@pbs.edu.pl; Tel.: +52-340-86-79

Abstract: This paper discusses changes in the microstructure and water absorption of thin-coat mineral-based plasters after prolonged exposure to the external environment and infected with biological corrosion. The results of laboratory and field tests for external thermal insulation composite system Styrofoam-based plasters are presented. The test samples were taken after 6 years of exposure to the external environment. The microstructure parameters such as porosity distribution and water absorption of the plasters were evaluated. The pore size ranges that were sensitive to frost corrosion and capillary flow were separated in the porosity distribution. Based on the porosity and absorption changes, it was found that biological corrosion interferes with the microstructure of the thin-coat mineral-based plasters on the expanded polystyrene substrate.

Keywords: biocorrosion; ETICS system; facade surface

Citation: Dybowska-Józefiak, M.; Wesołowska, M. The Influence of Biofilm on Selected Properties of Thin-Coat Mineral-Based Plasters on EPS Substrate. *Materials* **2022**, *15*, 5963. https://doi.org/10.3390/ma15175963

Academic Editor: F. Pacheco Torgal

Received: 4 August 2022
Accepted: 23 August 2022
Published: 29 August 2022

1. Introduction

Beneficial long-term properties combined with excellent protection against blowing rain and high thermal insulation quality are the reasons why ETICSs (external thermal insulation composite systems) [1–3] have become so popular in Central Europe. The multilayer exterior coating is evaluated as a less vulnerable solution in comparison with traditional plaster coatings [4–6]. However, earlier ETICS applications revealed particularly poor resistance to mechanical damage and biological infestation of plasters [7]. Changes to the exterior wall's finished surface due to microbial growth are generally accepted as 'patina' as long as they are uniformly distributed, while local contamination or algae concentrations are often evaluated as a 'visually adverse' [8] (Figure 1). However, the primary cause of plaster biocorrosion is microorganism presence.

Figure 1. Changes in the exterior wall's finished surface (author's archives).

Bacteria and fungi are involved in the degradation of any natural material. It has been found that from 20 to 30% of all damage is caused by microorganisms [5,9]. The process of biological deterioration of materials is referred to as biodeterioration, biocorrosion, or biological corrosion and is closely related to the organism metabolism [9]. Non-organic materials are exposed to corrosion with acidic metabolites produced by microorganisms (e.g., oxalic acid, gluconic acid, and citric acid) that form ion compositions with, e.g., lime and magnesium ions, taken from the material, which contributes to structure weakening [9,10]. Microorganisms are deposited onto surfaces to form clusters, surrounded by a protective layer of mucus, which are called biofilms or biological films (Figure 2a,b). Different biofilm areas are differentiated with density, organic matter availability, and oxygen content. Consequently, organisms in the film are characterized with different metabolic activity or even no activity at all [7].

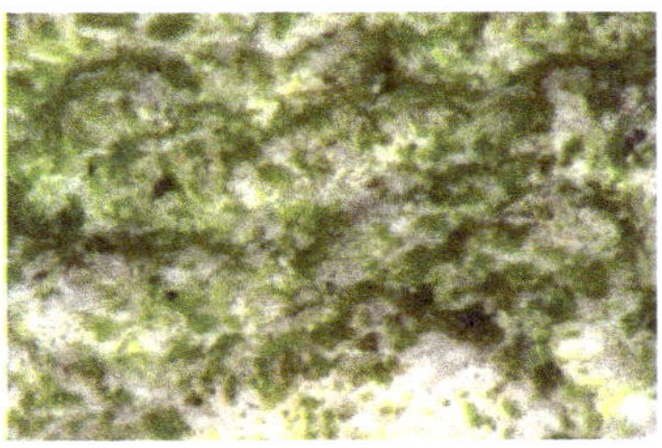

(a)

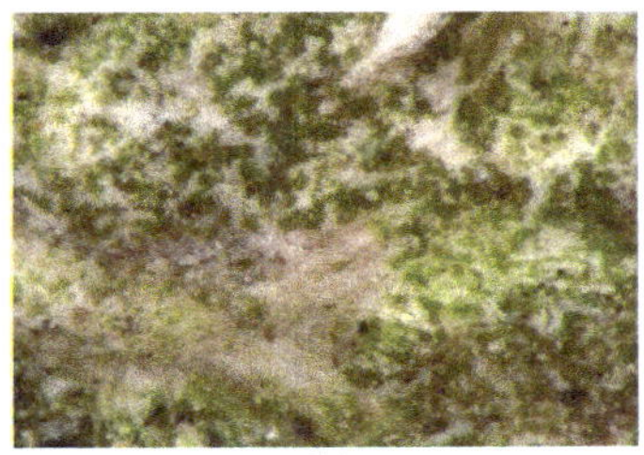

(b)

Figure 2. The surface of the (**a**) northern mineral-based dashed plaster—200× magnification photo taken with KEYENCE digital microscope (VHX-7000); (**b**) northern mineral-based filled plaster—200× magnification photo taken with KEYENCE digital microscope (VHX-7000).

Such behavior minimizes the diffusion of unfavorable substances from the outside (e.g., biocides) and enhances colony survival [10]. The biofilm affects changes in the appearance and structure of the engineering material [9]. It causes changes in porosity, is associated with vapor diffusion inside the material, absorbs contaminants, and leads to a change in the aerobic conditions into anaerobic conditions in the occupied area [7]. Acids, produced as a by-product of respiration and photosynthesis of biofilm-forming organisms, are often the cause of exterior wall finish deterioration [9]. This form of biodegradation is presented by bacteria, e.g., sulfur bacteria, that oxidize sulfur-containing nutrients to sulfuric acid and decompose scale or nitrifying bacteria with the production of corrosive nitric acid [10]. In case of building structures, the term biodeterioration can often be found, which means a general loss in the quality of building materials as a consequence of biological factors [10].

The study by Dylla et al. [11] on infected plaster surfaces of the ETICS system on selected structures in Bydgoszcz (Poland) found no bacteria and cyanobacteria that are often a component of aerophytic populations. However, three taxa of algae from the Chlorophyceae were successfully determined. By far, the dominant species was Apatococcus vulgaris Brand emend, Geitler—a cosmopolitan, aerophytic taxon with broad ecological valence. Filamentous green algae thrived in the wettest areas: Chlorohormidium flaccidum (Kütz.) Fott var. falaccidum Braun and the typically aquatic Ulothrix oscillarina (Kütz.) that is quite resistant to desiccation.

Changes in porosity and pore structure take place along with abiotic factors. According to a study conducted by Bochen, Gil, and Szwabowski, the average pore size decreases over time, but the open porosity volume increases [12]. In turn, water absorption and total porosity play key roles in microorganism growth on the surface of building materials, both of which are directly responsible for water and nutrient retention on and inside the samples [13].

Neville [14] and other researchers [8] believe that material structure tightness is essential for water and gas penetration. These parameters significantly affect strength, density, sorptivity, capillary suction, absorbability, permeability, and freeze–thaw durability. Therefore, they are used in methods for predicting the durability of building materials [12]. Water absorption and retention in materials are controlled by total porosity and pore size distribution (Tran at al. [15]). The effect of porosity and its structure on the physical properties of building materials was confirmed by Neville [14], Fagerlund [8], and other researchers [16–22], who conducted studies on concrete, mortar, cement paste, ceramic stone, gravel, and other materials. It is well known that a smaller pore volume has a good effect on the properties of cementitious materials since the tightness of material structure is of fundamental importance in terms of the penetration effected by water and gases. These parameters substantially influence the strength, density, sorptivity, capillary suction, absorption, permeability, and frost resistance, which are applied in the durability prediction methods of building materials [12,17,19,20,22]. The hardening process, which can take more than a year, can also increase the number of micropores and mesopores of less than 100 nm in size and reduce the number of capillary pores, which improves the material tightness. This promotes less penetration of moisture, and SO_2 and NO_x aggressive gases [23] into the microstructure of plasters and thus improves the material durability. For this reason, the types and size of pores are quite important, especially open pores that allow for penetration of liquids and gases. According to Neville [14], a large number of pores less than 5 nm in size hinders water penetration. In pore sizes of 5–100 nm, water can flow due to diffusion. In pore sizes larger than 100 nm, water can flow due to capillary adsorption. In addition, at low temperatures in pores of 100–1000 nm in size, freezing water causes the largest damage in material [8]. With the expansion of areas containing capillary water, a more intense capillary flow occurs [24]. According to [25], capillary absorption and transport of water are only possible in pores in the range from 100 nm to 100,000 nm in size, which are also called capillary pores.

The aim of this study is to evaluate the changes in the microstructure of mineral thin-coat plasters after long-term exposure to environmental conditions and to determine the effect of biofilm on the degradation of these plasters.

2. Test Material

Test materials were mineral-based dashed plasters of the ETICS system. The study was performed on three sample groups:

- Initial laboratory samples (designated as group A);
- Initial laboratory samples after freeze–thaw cycles (designated as group B);
- Samples taken from the test bed subjected to a natural ageing process (designated as group C).

Samples for laboratory tests (Figure 3) were prepared according to ETICS system specification on a 2 cm thick Styrofoam insulation layer with dimensions of 75 × 300 mm that ensure the required surface area of 225 cm^2. The study was performed on a group of 2 × 6 samples conditioned for 28 days in laboratory conditions.

Figure 3. Plaster samples for laboratory tests: (**a**) sample conditioning method; (**b**) shape and sample texture (filled plaster).

The samples to be tested after freeze–thaw cycles were dried to a constant weight at +40 ± 2 °C. They were then soaked in water according to the procedure described in Section 3.1. The samples were placed in a climate chamber that was programmed with the following cycles:

- Heating up to 60 °C at 80% humidity within 1 h and maintaining constant temperature and humidity for 3 h;
- Reducing temperature to –20 °C within 1 h and maintaining it for 3 h.

After three complete cycles (24 h), the samples were weighed, and the humidity level of the samples was supplemented by soaking them in water according to the procedure described in Section 3.1 and weighed again. All 81 aging cycles were conducted this way. The number of cycles was derived from an analysis of temperature transitions through 0 °C during the period of test stand operation.

The test site was located on the campus of the University of Technology in Bydgoszcz, Poland. The site consists of four walls made of aerated concrete, 24 cm thick, peripherally insulated with 15 cm thick EPS70 Styrofoam. The surfaces of the plaster coatings were oriented on the test bench with respect to all directions in order to determine the influence of this factor as a determinant of the aesthetics of the plaster coatings. The protective function in ETICS systems is mainly performed by structural plaster. They should be characterized by the following properties: low absorbency, diffusivity, resistance to dirt, resistance to biological infestation, and UV resistance [14,15]. Therefore, four different layers of plaster were laid on each surface of the study area, separated by expansion joint profiles: acrylic, mineral, silicate, and silicone. The plasters were outdoor exposed for 6 years (2016–2022).

The test sample was obtained with the quartering method. To conduct the tests, samples of 10 cm × 10 cm were taken from the site, from the north and south elevations. Sampling began by cutting the plaster surface with an angle grinder and then using a knife to cut out the fragments along with the thermal insulation layer (Figure 4).

(**a**)

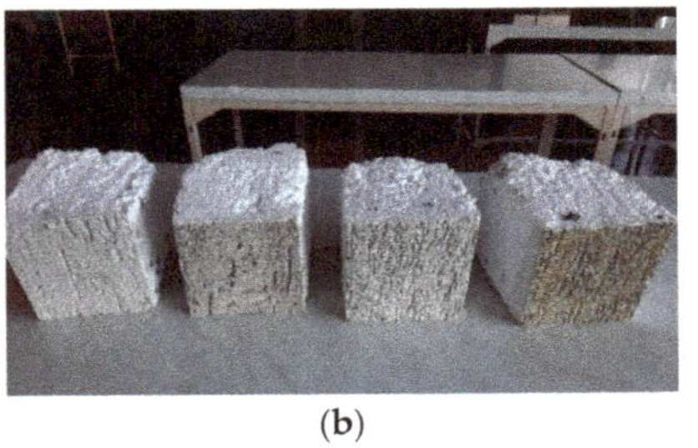

(**b**)

Figure 4. Sampling at the field site: (**a**) cutting the plaster surface; (**b**) samples that were cut out.

After sampling, a fragment of Styrofoam was cut off, leaving a 2 cm thick layer. The type C samples prepared this way were dried to a constant weight at +40 ± 2 °C.

The order of testing was based on the standard size of samples taken for testing, which were fragmented at a later stage. In the first stage, the following tests were performed: water absorption and freeze–thaw cycles for group B samples.

Within each group of samples, material was collected for microstructure determination. To ensure the proper representation of the results, material was taken from each of the 6 samples and then combined. The test sample was obtained with the quartering method.

3. Test Methods

3.1. Water Absorption Test

The water absorption for plasters was determined according to EN ISO 15148 [26]. Three samples were prepared for each plaster type, each with an area >100 cm^2. The lateral surface of each sample was protected with silicone. The samples were then dried to a constant weight and placed in a water-filled vessel equipped with:

- A system to maintain a constant water level with an accuracy of ±2 mm;
- A load to hold the sample in a particular position;
- Spacers to hold the sample at least 5 mm from the bottom (Figure 5).

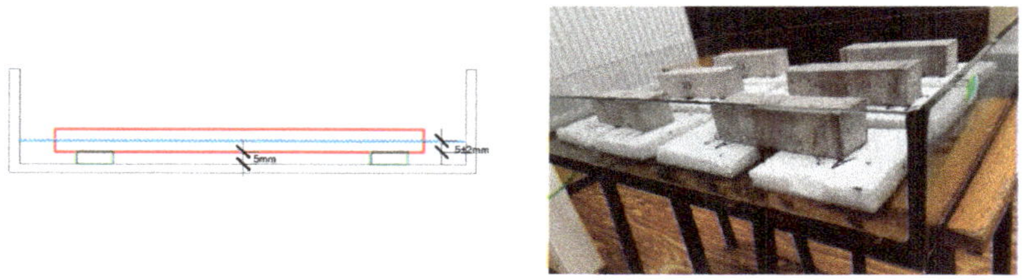

(**a**) (**b**)

Figure 5. Test stand for tests of plaster water absorption: (**a**) schematic diagram; (**b**) view.

The first mass measurement was made after 5 min: the sample was taken out of the water, dried with a damp sponge, and then weighed with an accuracy of ±0.1% of its mass (Figure 6). The procedure of immersion, removal, drying, and weighing was repeated after 20 min, 1 h, 2 h, 4 h, 8 h, 12 h, and 24 h. For samples taken from the field site, the test was performed in two variants: with biofilm and after cleaning the plaster surface with biocide. The infected surface was pre-treated to remove bloom before the application of the biocide. The samples were evenly sprayed with biocide and left for 12 h. The contamination was then removed mechanically using high-pressure water. The preparation was applied twice with a 12 h interval.

(**a**) (**b**)

Figure 6. Test site for natural aging test: (**a**) 2016; (**b**) 2022.

The water absorption coefficient was determined based on the following formulas:

$$A_w = \frac{\Delta m'_{tf} - \Delta m'_0}{\sqrt{t_f}} \tag{1}$$

where:

$\Delta m'_{tf}$—a value of Δm read from the timeline t_f, in kg/m^2;
$\Delta m'_0$—a value of Δm read from the timeline t_f for $t_f = 0$, in kg/m^2;
t_f—test time, in seconds.

$$W_w = \frac{\Delta m'_{tf} - \Delta m'_0}{\sqrt{t_f}} \tag{2}$$

where

$\Delta m'_{tf}$—a value of Δm read from the timeline t_f, in kg/m^2;
$\Delta m'_0$—a value of Δm read from the timeline t_f for $t_f = 0$, in kg/m^2;
t_f—test time, in hours.

3.2. Microstructure Test Using Mercury Intrusion Porosimetry (MIP)

The microstructure test was performed using a 9500 series AutoPore IV mercury intrusion porosimeter equipped with two ports: low and high pressures with a maximum value of 33,000 psi (228 MPa), which allows for measurements in the range of meso- and macropores. A penetrometer designed for granular and dusty material was used during test phase. Before conducting tests, the penetrometer was calibrated to determine the volume, compressibility, and thermal effect of the penetrometer used. An equilibrium time of 30 s was determined based on control measurements. As a result of the MIP test, the following parameters of the structure in question were determined: total pore volume, its skeletal density in mercury, and the distribution of the pore volume as a function of the pore diameter as an integral and differential relation.

The share of pore volume was calculated based on the following formulae:

- Pores responsible for freeze–thaw durability:

$$U_{frost} = \frac{\sum_{i=100\text{ nm}}^{1000\text{ nm}} IV_{frost}}{TIV} \cdot P \quad (3)$$

- Pores responsible for capillary transport:

$$U_{cap} = \frac{\sum_{i=100\text{ nm}}^{100{,}000\text{ nm}} IV_{cap}}{TIV} \cdot P \quad (4)$$

where

IV_{frost}—pore volume in the diameter range from 100 to 1000 nm;
IV_{cap}—pore volume in the diameter range from 100 to 100,000 nm;
TIV—total mercury intrusion;
P—total porosity.

4. Test Results

4.1. Water Absorption

Symptoms of biological corrosion occurred after two years, with varying degrees of severity depending on the orientation relative to the world directions (Figure 6).

The absorption results of tested samples are presented Table 1.

Table 1. Water absorption coefficient of mineral-based plaster.

Sample Type	Orientation	Water Absorption Coefficient	
		A_w (kg/(m$^2\times$s$^{0.5}$))	W_w (kg/(m$^2\times$h$^{0.5}$))
Initial (A)	-	0.0023	0.1372
After freeze–thaw cycles (B)	-	0.0011	0.0656
Taken from the field site (C)	S	0.0007	0.0433
with biofilm	N	0.0008	0.0466
Taken from the field site (C)	S	0.0016	0.0969
after biocide cleaning	N	0.0036	0.2177

The water absorption coefficient A_w decreased twofold after the freezing cycles (type B samples) compared with the output samples (type A samples)—the value of A_w coefficient decreased from 0.0023 kg/(m$^2\times$s$^{0.5}$) to 0.0011 kg/(m$^2\times$s$^{0.5}$).

The field site samples where the biofilm layer is located show a lower water absorption coefficient than the initial samples $A_w = 0.0023$ kg/($m^2 \times s^{0.5}$), with negligible differences between S and N orientation.

After sample cleaning and removing the biofilm as a result, the A_w index for the northern samples in S orientation is 0.0016 kg/($m^2 \times s^{0.5}$) and that in the N orientation is 0.0036 kg/($m^2 \times s^{0.5}$).

After biofilm removal, the southern orientation showed an increase in the A_w coefficient from 0.0007 kg/($m^2 \times s^{0.5}$) to 0.0016 kg/($m^2 \times s^{0.5}$)). In the northern orientation, however, the A_w coefficient increases from 0.0008 kg/($m^2 \times s^{0.5}$) to 0.0036 kg/($m^2 \times s^{0.5}$) (Table 1).

Large dynamic of water absorption in samples with biofilm removed can be observed in the N orientation from the beginning of the study (Figure 7); these changes are three times greater than the changes in the sample before cleaning. This relationship is not observed in the S orientation.

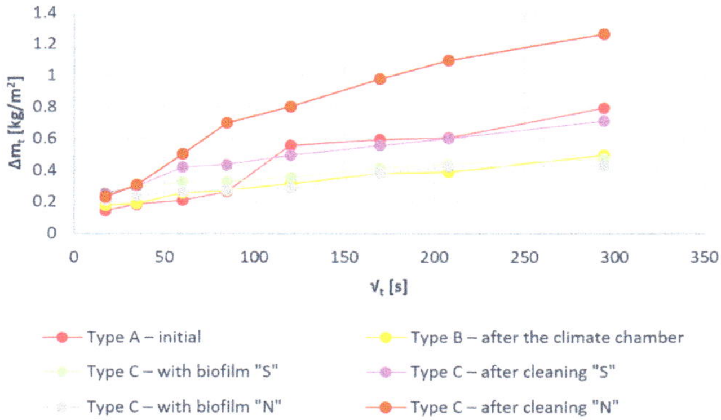

Figure 7. Graphs of $\Delta m_t/F$ as a function of the time element $t^{0.5}$ related to mineral-based plaster samples.

4.2. Microstructure

The microstructure results are shown in Figures 8 and 9 and in Tables 2 and 3.

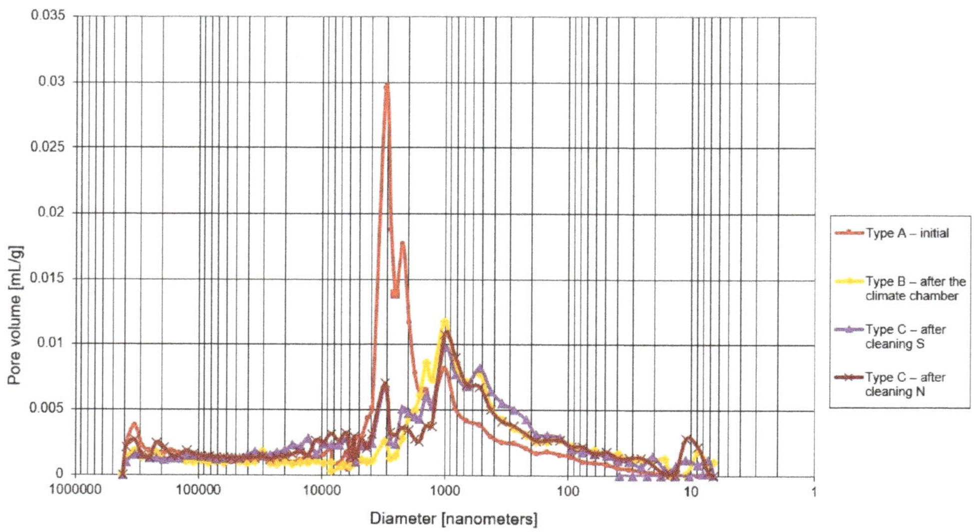

Figure 8. Differential curves of pore distribution for mineral-based plaster.

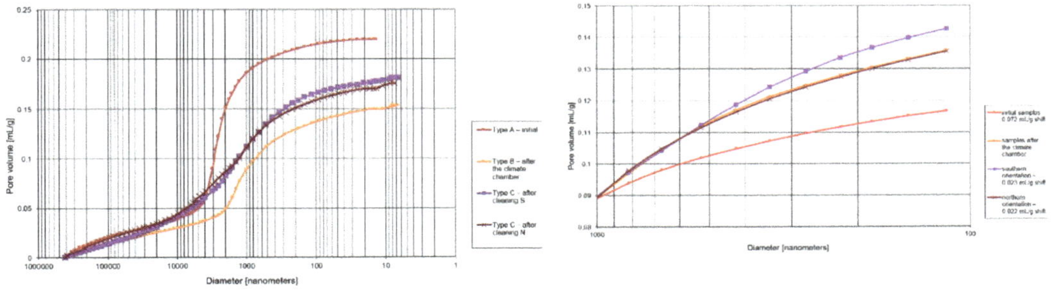

(**a**) (**b**)

Figure 9. Integral curves of pore distribution for mineral-based plaster: (**a**) full measuring range; (**b**) pore range related to freeze–thaw durability (from 100 nm to 1000 nm).

Table 2. Porosity test results.

Sample Type	General Porosity (%)	U Cap (%) (100 nm to 100,000 nm)	U Frost (%) (100 nm to 1000 nm)	Dominant Diameters (nm)
Initial (A)	34.2	30.0	4.4	3000
After freeze–thaw cycles (B)	26.8	21.2	8.2	1000
Taken from the field site (C), S orientation	29.5	24.8	9.0	1000
Taken from the field site (C), N orientation	29.2	22.8	7.7	1000 3000

Table 3. Mercury intrusion for defined pore ranges related to capillary transport and frost corrosion.

Sample Type	Mercury Intrusion for a Given Range of Pores, mL/g		Volume Changes, mL/g	
	IV_{cap}	IV_{frost}	IV_{cap}	IV_{frost}
	100 nm ÷ 100,000 nm	100 nm ÷ 1000 nm	100 nm ÷ 100,000 nm	100 nm ÷ 1000 nm
Initial (A)	0.192562	0.02791	–	–
After freeze–thaw cycles (B)	0.121733	0.046944	36.8	−68.2
Taken from the field site (C), S orientation	0.152485	0.054965	20.8	−96.9
Taken from the field site (C), N orientation	0.137783	0.046643	28.4	−67.1

Total porosity for the initial plaster samples was 34.2%. As the samples were subjected to specific freezing cycles, the porosity decreased to 26.8%. Field site samples show lower porosity than the initial samples. The porosity for the northern orientation is 29.2%, and the porosity for the southern orientation is 29.5%.

The volume share of pores related to capillary transport (from 100 nm to 100,000 nm) for the initial samples was 30.0%. As a result of freeze–thaw cycles in laboratory conditions, the value decreased to 21.2%. The utilization of plasters under environmental conditions resulted in a decrease in the value of the initial pore share for the S orientation to 24.8% and for the N orientation to 22.8%.

The pore share related to freeze–thaw durability (100 nm to 1000 nm) for the initial samples was 4.4%. After the freeze–thaw cycles in the laboratory, it increased to 8.8%. In case of the field site in the S orientation, it is 9.0% and in N orientation, it is 7.7%. Based on the analysis of the integral curves in the porosity range of 100–1000 nm (Figure 10), it was found that the pore volume in the dimension range of 300–1000 nm increased when compared with the initial sample.

Figure 10. Photographs of the plaster surface with biofilm and after cleaning for the N and S orientation: (**a1**) N-oriented sample with biofilm; (**b1**) N-oriented sample after cleaning; (**a2**) S-oriented sample with biofilm; and (**b2**) S-oriented sample after cleaning.

When analyzing the mercury intrusion in mL/g for the specific pore ranges (Table 3), it can be concluded that the direct volume changes are very large. For the pore range related to capillary transport (from 100 nm to 100,000 nm), the largest changes were observed for laboratory freeze–thaw cycles—volume reduction of 36.8%—and the smallest for the south-oriented field site station—volume reduction of 20.8%. For the pore range related to freeze–thaw durability, similar results were obtained for freeze–thaw cycles and for the field site station in the N orientation—volume increases of more than 65%. Samples taken from the S-oriented field site station show almost twofold increase in mercury intrusion.

5. Discussion

Comparing the northern and southern orientations, it can be seen that the north-oriented plaster features higher water absorption than the south-oriented plaster. This relationship is visible in both non-cleaned samples as well as in samples with biofilm removed.

On the other hand, when comparing field site samples with biofilm and samples subjected to freeze–thaw cycles, it can be observed that field site samples cut from the station have lower Aw coefficient than samples subjected to freeze–thaw cycles. On the other hand, after cleaning them, the Aw coefficient increases to 0.0016 $kg/(m^2 \times s^{0.5})$ in the case of the southern sample and to 0.0036 $kg/(m^2 \times s^{0.5})$ in the case of the northern sample, and at the same time, it is higher than the Aw coefficient for the samples tested in the climate chamber, where Aw = 0.00116 $kg/(m^2 \times s^{0.5})$.

Based on microscopic images, it was determined that the removal of biofilm from the N-oriented plaster compromises the plaster surface and texture (Figure 10).

The initial samples are featured by the lowest share of pores related to the freeze–thaw durability; these pores constitute 12.9% of total porosity. On the other hand, the initial samples show the highest share of pores related to capillary transport—87.7% of the total porosity. The dominant diameter of 3000 nm, determined from the differential pore size distribution, is within the range of capillary pores but outside the range of pores related to their freeze–thaw durability.

For the samples subjected to freeze–thaw cycles, the pore share in relation to total porosity for freeze–thaw durability is 30.6%, and the pore share for capillary transport is 79.1%. As a result of freeze–thaw cycles, the dominant diameter of 3000 nm disappeared, but a new dominant diameter of 1000 nm appeared in the pore region related to capillary transport.

Compared with the samples subjected to freeze–thaw cycles, the plaster samples taken from the field site station feature a higher pore share related to capillary transport: for the N orientation, by 78.1% and, for the S orientation, by 84.4%. On the other hand, the pores related to freeze–thaw durability for the field site represent 26.4% of the total porosity for the N orientation and 30.5% for the S orientation. The porosity distribution of south-oriented plaster is similar to the plaster subjected to freeze–thaw cycles. In case of the northern orientation, a bimodal porosity distribution takes place, where the dominant diameter of 3000 nm persists and, in addition, a second dominant diameter of 1000 nm is formed in the pore range related to capillary transport.

6. Conclusions

This paper presents the results of the study of thin-layer mineral-based plasters with a microorganism biofilm that colonizes exterior wall finishes. For comparison purposes, initial samples, samples subjected to laboratory freeze–thaw cycles, and samples from a south-oriented and north-oriented field test station were tested. On the field site station, the biofilm visible to the unaided eye appeared already after 2 years and developed for further 4 years. The number of freeze–thaw cycles in the laboratory test corresponded to the recorded number of passes through 0 °C over a 6-year period.

As a result of the study, changes in the absorption and the porosity structure of the thin-coat mineral-based plasters taken from the field site as well as in the samples subjected to freeze–thaw cycles were found.

The test for water absorption changes showed a correlation with mercury intrusion and porosity changes in the capillary range in the case of cyclic freeze–thaw and a southern orientation of the field site station. For the northern orientation of the field site station, no such correlation was found—after biofilm removal with biocides, there are very high dynamics of water absorption during the test period (24 h) and, consequently, a very high coefficient of water absorption at similar porosity. This is probably the result of intensive biofilm growth and removal treatment procedures that disturb the surface and texture of the thin-coat mineral-based plaster.

The literature details the porosity range related to the freeze–thaw durability of plasters from 100 nm to 1000 nm. During the course of the study, it was found that both cyclic freeze–thaw and the utilization of plaster in a natural environment introduced changes in a narrower range of diameters, that is from 300 to 1000 nm. A common dominant diameter of 1000 nm also appeared in this range. Therefore, it is suggested that the porosity range responsible for the freeze–thaw durability of thin-coat mineral-based plasters should be narrowed from 300 nm to 1000 nm.

Further studies are expected in order to analyze the change in composition of the plasters due to freeze–thaw cycles and further development of biofilm on the plasters.

Author Contributions: Conceptualization, M.W.; methodology, M.D.-J. and M.W.; validation, M.D.-J. and M.W.; formal analysis, M.D.-J. and M.W.; investigation, M.D.-J.; resources, M.D.-J. and M.W.; data curation, M.D.-J. and M.W.; writing—original draft preparation, M.D-J.; writing—review and editing, M.W.; supervision, M.W. All authors have read and agreed to the published version of the manuscript.

Funding: This research received no external funding.

Institutional Review Board Statement: Not applicable.

Informed Consent Statement: Not applicable.

Data Availability Statement: Other results are not published.

Conflicts of Interest: The authors declare no conflict of interest.

References

1. EAD 040083-00-0404 External Thermal Insulation Composite Systems (ETICS) with Renderings. Available online: https://www.nlfnorm.cz/en/ehn/6669 (accessed on 6 June 2022).
2. ETAG 004 External Thermal Insulation Composite Systems (ETICS) with Renderings. Available online: https://www.nlfnorm.cz/en/ehn/6705 (accessed on 6 June 2022).
3. Parracha, J.L.; Borsoi, G.; Flores-Colen, I.; Veiga, R.; Nunes, L.; Dionisio, A.; Gomes, M.; Faria, P. Performance parameters of ETICS: Correlating water resistance, bio-susceptibility and surface properties. *Constr. Build. Mater.* **2021**, *272*, 121956–121970. [CrossRef]
4. Kukletova, I.; Chromkova, I. Testing of biocidal properties of thermal insulation system during material life cycle. In *IOP Conference Series: Materials Science and Engineering*; IOP Publishing: Bristol, UK, 2021; Volume 1205.
5. Michałowski, B.; Marcinek, M.; Tomaszewska, J.; Czernik, S.; Piasecki, M.; Geryło, R.; Michalak, J. Influence of Rendering Type on the Environmental Characteristics of Expanded Polystyrene-Based External Thermal Insulation Composite System. *Buildings* **2020**, *10*, 47. [CrossRef]
6. Künzel, H.; Künzel, H.M.; Sedlbauer, K. Long-Term Performance of External Thermal Insulation Systems (ETICS). *Acta Sci. Pol. Archit.* **2006**, *1*, 11–24.
7. Johansson, S. Biological Growth on Rendered Facades. Ph.D. Thesis, Lund University, Division of Building Materials, Lund, Sweden, 2011.
8. Fagerlund, G. *Durability of Concrete Structures*; Arkady: Warsaw, Poland, 1997.
9. Stanaszek-Tomal, E. The Problem of Biological Destruction of Façades of Insulated Buildings—Causes and Effects. In *IOP Conference Series: Materials Science and Engineering*; IOP Publishing: Bristol, UK, 2017; Volume 245.
10. Wiejak, A.; Miklaszewska, J. Odporność na glony tynków cienkowarstwowych do stosowania na zewnątrz obiektów. *Mater. Bud.* **2012**, *481*, 28–29.
11. Dylla, A.; Paczuska, B.; Wernerowska-Frąckiewicz, Z. Wstępne badania nad glonami aerofitycznymi porastającymi elewację budynków w Bydgoszczy ocieplonych metodą lekką-mokrą. *Ochrona Przed Korozją* **2008**, *5*, 47–50.
12. Bochen, J.; Gil, S.; Szwabowski, J. Influence of ageing process on porosity changes of the external plasters. *Cem. Concr. Compos.* **2005**, *27*, 769–775. [CrossRef]
13. Wilimzig, M. Biodeterioration od building materials. In Proceedings of the 8th International Congress on Deterioration and Conservation of Stone, Berlin/Heidelberg, Germany, 30 September–4 October 1996; Volume 18, pp. 579–583.
14. Neville, A.M. *Properties of Concrete*, 4th ed.; Arkady: Cracow, Poland, 2000.
15. Tran, T.; Govin, A.; Guyonnet, R.; Grosseau, P.; Lors, C.; Damidot, D.; Deves, O.; Ruot, B. Influence of the intrinsic characteristics of mortars on their biofouling by pigmented organisms: Comparison between laboratory and field scale experiments. *Int. Biodeterior. Biodegrad.* **2014**, *86*, 334–342. [CrossRef]
16. Tada, S. Microstructural approach to frost resistance of highly porous materials. *Durab. Build. Mater. Compon.* **1996**, *1*, 299–308.
17. Zhang, B. Relationship between pore structure and mechanical properties of ordinary concrete under bending fatigue. *Cem. Concr. Res.* **1998**, *28*, 699–711. [CrossRef]
18. Kearsley, E.P.; Wainwright, P.J. The effect of porosity on the strength of foamed concrete. *Cem. Concr. Res.* **2002**, *32*, 233–239. [CrossRef]
19. Kumar, R.; Bhattacharjee, B. Porosity, pore size distribution and strength of concrete. *Cem. Concr. Res.* **2003**, *33*, 155–164. [CrossRef]
20. Zhihua, P.; Dongxu, L.; Jian, Y.; Yang, N. Properties and microstructure of the hardened alkali-activated red mud-slag cementitious material. *Cem. Concr. Res.* **2003**, *33*, 1437–1441.
21. Moropoulou, A.; Polikreti, K.; Bakolas, A.; Michailidis, P. Correlation of physicochemical and mechanical properties of historical mortars and classification by multi variate statistics. *Cem. Concr. Res.* **2003**, *33*, 891–898. [CrossRef]
22. Matusinović, T.; Šipušić, J.; Vrbos, N. Porosity–strength relation in calcium aluminate cement pastes. *Cem. Concr. Res.* **2003**, *33*, 1801–1806. [CrossRef]
23. Martinez-Ramirez, S.; Puertas, F.; Blanco-Varela, M. Studies on degradation of lime mortars in atmospheric simulation chambers. *Cem. Concr. Res.* **1997**, *27*, 777–784. [CrossRef]
24. Barberousse, H.; Brayner, R.; Do Rego, A.M.B.; Castaing, J.-C.; Beurdeley-Saudou, P.; Colombet, J.-F. Adhesion of facade coating colonisers, as mediated by physico-chemical properties. *Biofouling* **2007**, *23*, 15–24. [CrossRef] [PubMed]
25. Orazio, M.; Cursio, L.; Graziani, L. Effects of water absorption and surface roughness on the bioreceptivity of ETICS compared to clay bricks. *Build. Environ.* **2014**, *77*, 20–28. [CrossRef]
26. *EN ISO 15148:2004*; Hygrothermal Performance of Building Materials and Products—Determination of Water Absorption Coefficient by Partial Immersion. ISO: London, UK, 2004.

materials

Article

Evaluation of Frost Impact on Traditional Ceramic Building Materials Utilized in Facing Walls

Anna Kaczmarek * and Maria Wesołowska

Faculty of Civil and Environmental Engineering and Architecture, Bydgoszcz University of Science and Technology, 85-796 Bydgoszcz, Poland
* Correspondence: anna.kaczmarek@pbs.edu.pl

Abstract: This paper takes into consideration the performance of traditional bricks as part of a building exterior wall finish. Exterior wall materials change their properties when exposed to external environment. This process is extended over time and its intensity is closely related to microstructure, moisture and freeze–thaw cycles. Two methods of freeze–thaw durability tests were used in this study: standard and defined by the authors. The authors' method incorporated the actual conditions of masonry unit function in exterior wall finish, i.e., cyclical effects of precipitation water, changes in temperature and air humidity. The laboratory test study included 50 freeze–thaw cycles. Three characteristic ranges of pore dimensions were indicated in the analysis: below 0.1 μm, between 0.1 and 3.0 μm and above 3 μm. Based on the method of freeze–thaw durability testing, the areas of microstructure changes were determined. The obtained results were related to the absorption of ceramic building materials. The authors' method confirms the usage of traditional ceramic building materials designed for use in protected walls against water penetration in unprotected exterior wall finish. The critical water saturation method of masonry units (standard) based on extreme environmental conditions generates significant changes in porosity distribution that do not reflect real, i.e., moderate, conditions. This method is appropriate for masonry units operating in severe conditions, i.e., F2. The aim of this study is to suggest a methodology for durability tests of traditional ceramic masonry units to cyclic freezing and thawing, which are only exposed to F1 (moderate) conditions during operation. Changes in the microstructure of the ceramic building materials were used as the primary evaluation criterion. In order to determine the effect of cyclic temperature changes, the freeze–thaw durability test was performed according to generally accepted standard procedures and in-house methodology. The purpose of the study is to point out the individual approach for the analysis of the material–environment system. At the same time, it should inspire researchers to innovative methods which use external conditions in a laboratory environment.

Keywords: microstructure; freeze–thaw durability; absorption; traditional ceramic building materials

Citation: Kaczmarek, A.; Wesołowska, M. Evaluation of Frost Impact on Traditional Ceramic Building Materials Utilized in Facing Walls. *Materials* **2022**, *15*, 5653. https://doi.org/10.3390/ma15165653

Academic Editor: Krzysztof Schabowicz

Received: 8 July 2022
Accepted: 13 August 2022
Published: 17 August 2022

1. Introduction

There is a large number of buildings in Poland, the elevations of which are made of masonry units that do not meet modern frost requirements. At the time of their erection, the walls were protected against blowing rain, soil moisture, etc. By their very existence, the walls made of traditional ceramic building materials testify to their long-term and stable adaptation to the environment. Although they do not meet modern durability requirements according to microexposure classes (Table 1), their retention state is good. This is due to the applied elevation and moisture barrier solutions. These walls are made of ceramic units that, according to the recommendations of EN 771-1 [1], are intended to be used in solutions protected against water penetration. The solutions operate in conditions defined by the standard as moderate—F1—with exposure to moisture and freeze–thaw cycles. Therefore, they do not have the possibility of full water saturation, combined with frequent freeze–thaw cycles due to climatic conditions and the lack of adequate protection generating F2 severe conditions. Therefore, it is worth asking a question as to whether plastering is necessary after many years of exposure without the protection required by Eurocode 6 [2], or whether such situations can be treated as facing walls. This question is particularly relevant in the context of intensification of degradation processes caused, among other things, by rapid climate change (intense rain, short-term snow cover periods, multiple 0 °C passes, sun exposure). An important factor here is the flow of water film down the wall surface in rain, which in the case of intensive precipitation forms a uniform layer constituting the basic source of moisture.

Table 1. Classification of the microconditions of exposure of completed masonry [2].

Class	Micro Condition of the Masonry	Examples of Masonry in This Condition
MX 3.1	Exposed to moisture or wetting and freeze/thaw cycling but not exposed to external sources of significant levels of sulfates or aggressive chemicals.	Internal masonry exposed to high levels of water vapor, such as in a laundry. Masonry exterior walls sheltered by overhanging eaves or coping, not exposed to severe driving ram or frost. Masonry below frost zone in well drained non-aggressive soil. Exposed to freeze/thaw cycling.
MX 3.2	Exposed to severe wetting and freeze/thaw cycling but not exposed to external sources of significant levels of sulfates or aggressive chemicals.	Masonry not exposed to frost or aggressive chemicals. Location: in exterior walls with cappings or flush eaves; in parapets; in freestanding walls; in the ground; and under water. Exposed to freeze/thaw cycling.
MX 4	Exposed to saturated salt air, seawater or deicing salts.	Masonry in a coastal area. Masonry adjacent to roads that are salted during the winter.
MX 5	In an aggressive chemical environment.	Masonry in contact with natural soils or filled ground or groundwater, where moisture and significant levels of sulfates are present. Masonry in contact with highly acidic soils, contaminated ground or groundwater. Masonry near industrial areas where aggressive chemicals are airborne.

Important factors that cause the degradation of facing walls are freeze–thaw cycles of water and crystallization of soluble mineral salts [3–7]. In the case of subject literature, freeze–thaw durability is analyzed in two aspects [4,8,9]. The first group of tests focuses on macroscopic effects including weight loss, surface damage and strength parameters. The second group refers to the microstructure study strictly relating it to the resistance of ceramic building materials to frost corrosion [3,10–13].

In this case, pore dimensions are mentioned as an important parameter that focuses on three basic ranges:

- Large pores, larger than 3.0 μm, that feature a favorable effect on the freeze–thaw durability [8,9,14]. This is due to the fact that large pores are a kind of compensation chamber for the stresses during ice crystallization [15]. Moreover, the crystallization pressure is lower in larger pores [16].
- Medium-sized pores of 3.0 to 0.1 μm, which are considered to be critical pores that determine the freeze–thaw durability of the bricks [14,17].
- Small pores of less than 0.1 μm, where the water freezing point is well below 0 °C [18].

The range of critical pores was narrowed by Elert et al. [19] to the scope of 0.2 μm to 2.0 μm. Culturone et al. [20] also found that pores smaller than 2.0 μm in diameter are responsible for water absorption and retention increase. Consequently, they determine the freeze–thaw durability of ceramic building materials. According to Tang et al. [21], the destructive effect occurs in pores with diameters below 1 μm, which increase in size under the influence of cyclic freezing and defrosting processes and thus cause a shift in the porosity structure towards larger pores (1.0 to 5.0 μm). Koroth [3] and Kung [22] had similar observations. According to [23], the shape and distribution of pores significantly affect the durability of masonry units There is no unified approach to the freeze–thaw durability test in the mentioned studies. At least four testing methods are indicated [14]:

- Critical degree of water saturation of complete masonry units;
- Critical degree of water saturation of crushed masonry units;
- Exposure of masonry units to actual climatic conditions;
- Freezing and thawing with sprinkling irrigation of test panels in accordance with EN 772-22 [24].

The method for determining the freeze–thaw durability depends on the adopted methodology and the number of freezing and thawing cycles. Frost corrosion results not only in mechanical property changes, but also in the way the material behaves in contact with water. These changes directly affect the moisture condition of the products in the walls.

In most cases, facing walls made of traditional ceramic building materials operate in an environment defined as F1 (moderate). A review of the current state of knowledge does not identify an individual approach to the issue. Taking into account the standard requirements, masonry units made of traditional ceramic building materials do not pass the freeze–thaw durability test. The authors suggest distinguishing the methods of freeze–thaw durability testing for elements operated in severe conditions and in moderate conditions. The aim of this study is to suggest a methodology for the durability tests of traditional ceramic masonry units to cyclic freezing and thawing, which are only exposed to F1 (moderate) conditions during operation. Changes in the microstructure of the ceramic building materials were used as the primary evaluation criterion. In order to determine the effect of cyclic temperature changes, the freeze–thaw durability test was performed according to generally accepted standard procedures and in-house methodology.

2. Materials

The bricks analyzed as part of the study are dated from the 1980s. At that time in Poland, masonry units were sourced from local brickyards, which used Hoffmann ring furnaces fired with pulverized coal. The technology was based on the mechanical product molding, using a plastic method and their natural drying, followed by burning at a temperature of +980 °C for 216 h. Bricks were made from clay materials which originated from local deposits.

The tested products belong to the group of masonry units to be used in walls protected against water penetration [1], with dimensions 250 × 120 × 65 mm (Figure 1). Their quality parameters are summarized in Table 2.

Figure 1. A brick sample used for tests.

Table 2. Quality parameters of ceramic building materials produced from the local Pliocene clay deposit.

Quality Parameters	Range	Average Value
batched water	(17.2–41.8)%	29.67%
drying contraction	(4.0–13.0)%	9.2%
absorption (temp. 980 °C)	(7.4–10.3)%	8.8%
compressive strength (temp. 980 °C)	(16.0–24.0) MPa	19.5 MPa
content of granular marl with more than 0.5 mm fraction	(0.00–0.40)%	0.031%
efflorescence of sulfate salts	-	none or minimal irremovable bloom

The tests were carried out on masonry units manufactured 35 years ago, stored in laboratory conditions (temp. 20 ± 2 °C and air humidity 50 ± 5%) in 6 samples.

The tests were conducted on 3 sample groups:

- Output samples (A);
- Samples after the standard freeze–thaw durability test (B);
- Samples after the freeze–thaw durability test with authors' method (C).

The order of tests was adapted to the crashed stages of the samples. In order to ensure the representation of the study results, each masonry unit was divided into 4 sections. One section was taken from each masonry unit. This resulted in laboratory samples consisting of 6 parts taken from 6 masonry units (Figure 2). In the first stage, the absorption tests were carried out with output samples (A) as well as freezing and thawing cycles employed for the standard method (sample B) and the authors' method (sample C).

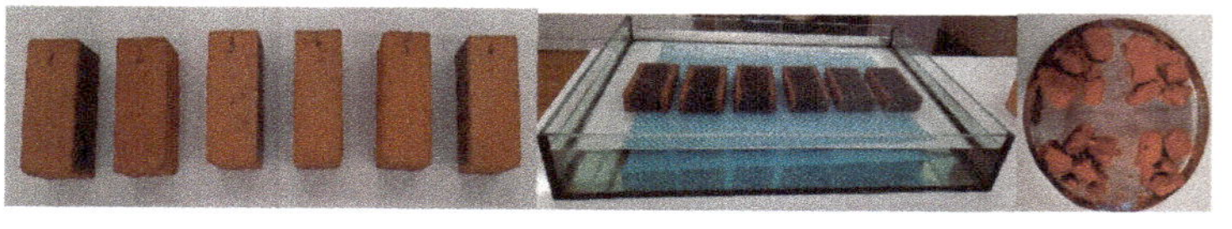

Figure 2. Preparation of samples for testing.

The samples were then used to determine the microstructure of each group (A, B and C). For this purpose, fragments of up to 1 cm^3 were taken from the facing surface of each sample. A laboratory sample was selected with the quartering method from the obtained material (Figure 2). Quartering consisted of coning the collected and thoroughly mixed material, then flattening and cross-dividing it into 4 parts. Two diagonal parts were removed and the remaining two parts were re-mixed and the selection process was repeated. This procedure was performed 3 times to obtain a laboratory sample volume of about 5 cm^3 that corresponds to the volume of the penetrometer tank.

3. Test Methods

3.1. Absorption Test

The absorption test was carried out with a soaking method of 6 brick samples. The samples were dried to a fixed weight at +40 °C, then weighed to the nearest 0.1%. The samples were placed in a vessel with supports of non-corrosive material. Then, they were flooded with water at room temperature up to $\frac{1}{2}$ of their height, after 3 h the water was replenished up to the level of $\frac{3}{4}$ of the sample height and after 3 h the water was replenished again until they were completely submerged (Figure 3).

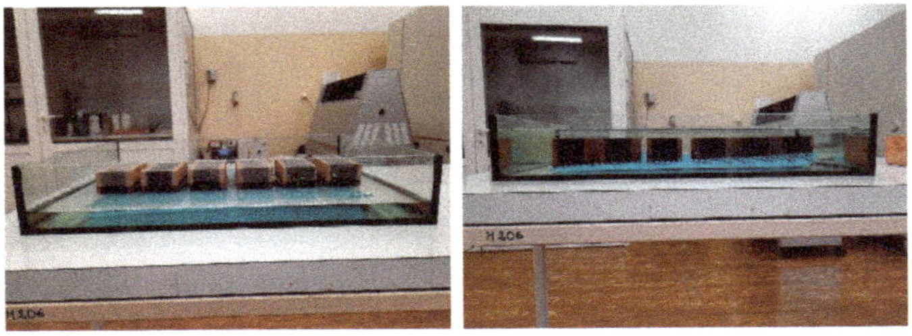

Figure 3. Absorption test.

The samples were kept in water until their fixed weight was established. Samples were taken out individually for weighting purposes (this protects them from drying out). External surfaces of the samples were wiped with a damp cloth.

Sample absorption n_m was calculated from the Formula (1), in %:

$$n_m = \frac{C_m - C_s}{C_s} \times 100\ [\%], \quad (1)$$

where:
C_m—weight of a water-soaked sample, g;
C_s—weight of dried sample, g.

3.2. Freezing and Thawing Cycles

Freezing and thawing cycles were performed with two methods:

- Critical degree of water saturation of masonry units;
- The authors' method.

In the first case, the samples were placed in a water container and saturated to a fixed weight. Then, the brick surfaces were dried from excess water and placed in a freeze–thaw durability test chamber (Figure 4). Fifty freeze–thaw cycles were assumed in the temperature range of −18 to +18 °C.

Figure 4. Sample conditioning in the authors' freeze–thaw durability method.

In the next step, the samples were placed in a climate chamber with the following operation cycles:

- The transition period of 0.5 h, including temperature adjustment from baseline to −18 °C;
- In total, 3.5 h at −18 °C;
- The transition period of 0.5 h, including temperature adjustment from −18 °C to +18 °C and humidity at 90%;
- In total, 3.5 h at +18 °C.

After each complete cycle, the samples were soaked again in water at +20 ± 2 °C according to the initial absorption procedure as described in EN 772-11 [24]. All 50 cycles were conducted this way.

3.3. Microstructure Test Using Mercury Intrusion Porosimetry

The microstructure test was performed using a 9500 series AutoPore IV mercury intrusion porosimeter equipped with two ports: low and high pressure with a maximum value of 33,000 psi (228 MPa), which allows measurements in the range of meso- and macropores. Before the actual test, the calibration and "blank test" of the penetrometer used in this test were carried out to determine volume, compressibility and thermal effect. An equilibrium time of 30 s was determined based on control measurements. As a result of the prepared sample test, the following parameters of the structure in question were determined: total pore volume, sample volume and its skeletal density, the distribution of the pore volume as a function of the pore diameter as integral and differential relation. There were 18 samples.

The share of the pore volume was calculated based on the formula:

$$U_{frost} = \frac{\sum_{i=0.1\mu m}^{3.0\mu m} IV_{frost}}{TIV} \cdot P, \quad (2)$$

$$U_{nondest} = \frac{\sum_{i>3.0\mu m} IV_{nondest}}{TIV} \cdot P, \quad (3)$$

$$U_{small} = \frac{\sum_{i<0.1\mu m} IV_{small}}{TIV} \cdot P, \quad (4)$$

where:

$IV_{nondest}$—% share of meso- and macropores larger than 3.0 μm in diameter;
IV_{frost}—% share of mesopores with diameters in the range of 0.1 to 3.0 μm;
IV_{small}—% share of nanopores smaller than 0.1 μm in diameter;
P—total porosity.

4. Test Results and Discussion

The results of the microstructure test indicate that, compared to the initial material, the samples tested in both with the standard and the authors' method obtained lower values of total porosity (Table 3).

Table 3. Tests results.

Test	Freezing and Thawing Cycles According to the Adopted Method		
	Output Sample	**Critical Degree of Water Saturation of Masonry Units**	**Authors' Method**
general porosity [%]	33.35	32.68	31.41
absorption [%]	13.84	14.55	14.34
dimension of dominant pores [μm]	1.3, 0.045	4.5	5.0

Lower value of porosity was obtained in the case of the material after treatment with the authors' method of freeze–thaw durability tests (–1.94%). In the case of the critical degree of water saturation of masonry units, the difference was –0.67%. The factor that essentially differentiates the features of the material covered in this study is the dominant pore diameter and porosity structure. The initial material exhibits a bimodal arrangement, typical for ceramic building materials, with dominant pores of 1.3 μm and 0.045 μm in diameter (Figure 5). In the analyzed cases, there is one dominant diameter, the value of which is much higher (for the samples after the authors' freeze–thaw durability test of 5 nm, and after the method of critical degree of water saturation of masonry units of 4.5 μm (Figure 5), respectively. The obtained dominant diameters above 3 μm are within the range to be considered safe from the point of view of frost damage. The differential curve plots of the pore size distribution of the tested brick samples are presented in Figure 6.

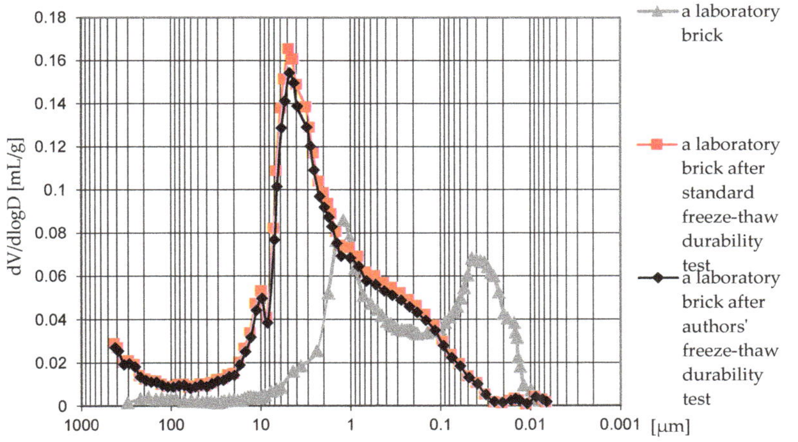

Figure 5. Differential curve of pore size distribution in test samples.

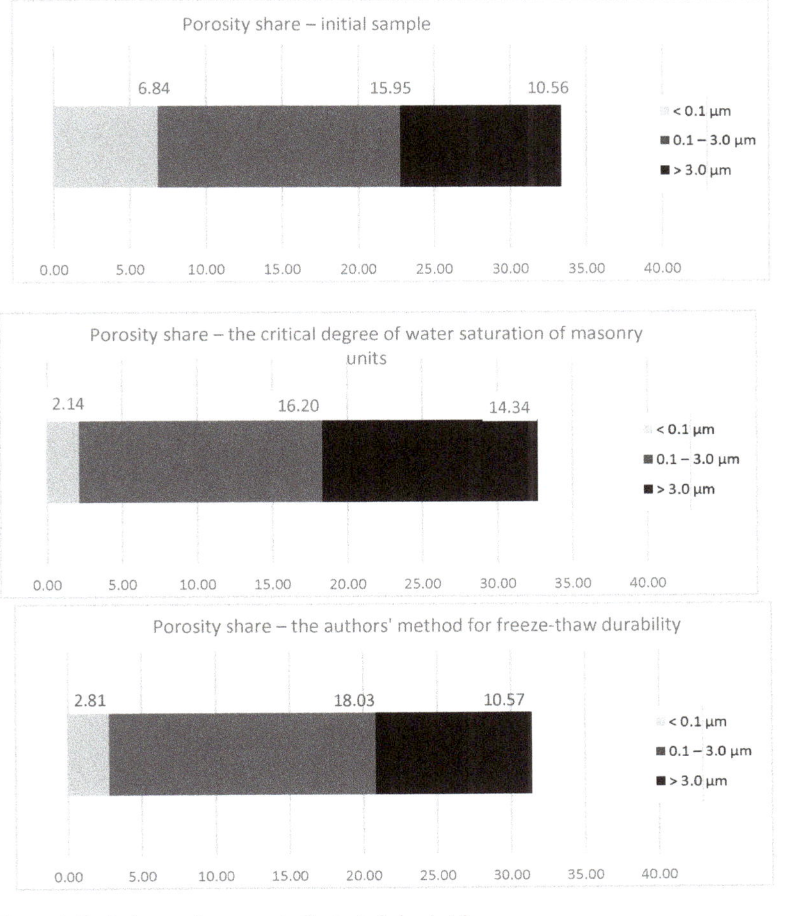

Figure 6. Typical porosity ranges in the tested clay bricks.

The form of frost damage is closely related to the pore distribution (Figure 6). In the initial material, the total porosity was 33.35% with a share of 15.95 percentage points from 0.1 to 3.0 μm in diameter. The remaining 10.56 percentage points were occupied with pores above 3 μm in diameter, and 6.84% with pores below 0.1 μm in diameter. In the samples that faced the authors' freeze–thaw durability test, the total porosity was 31.41% with the pore share in the range of 0.1–3.0 μm, increasing to 18.03 percentage points. The share of pores with diameters greater than 3.0 μm is maintained at the same level. There was a significant decrease in the share of pores below 0.1 μm at 2.81 percentage points. For the method of critical water saturation degree of masonry units, the total porosity was 32.68%, with a similar share of pores in the range of 0.1–3.0 μm (16.20 percentage points). There was a significant increase in the share of pores above 3.0 μm by 3 percentage points compared to the initial material. The share of pores below 0.1 μm decreased three times to 2.14 percentage points. In all analyzed cases, frost damage increased the share of porosity in the range of 0.1 to 3.0 μm. With cyclic freezing and thawing, it was found that pores below 0.1 μm were also responsible for the porosity structure changes. Frost influence damages them, resulting in increased pore size.

Porosity changes affect the moisture properties of the tested ceramic building materials. The absorption of the samples increases in relation to the initial material (Table 3) by about 1% regardless of the method of freezing and thawing cycles used.

5. Conclusions

This paper analyzes the influence of cyclic freezing and thawing on selected properties of masonry units intended to be used in walls protected against water penetration. Two test methods were adopted to determine the freeze–thaw durability of these ceramic masonry units. One of them is the commonly used standard method of the critical degree of water saturation of masonry units and the other is the proposed authors' method. The choice of the freeze–thaw durability test method should depend on the environmental conditions defined by the classes of microexposure and the intensity of the masonry unit exposure to moisture and freeze–thaw cycles. The standard test method—critical degree of water saturation of whole masonry units—is appropriate for severe conditions where it is possible. The authors' proposed method is appropriate for moderate conditions where only partial saturation of the masonry unit with water takes place. Facing walls without a protective layer of plaster in most cases are only exposed to moderate conditions (F1), for which the standard method of freeze–thaw durability test is inadequate. The authors suggest distinguishing the methods of freeze–thaw durability testing for masonry units used as face walls.

In the case of elements operating in severe conditions, it is suggested to use the standard method, while for elements operating in moderate conditions the authors' method described in the article is recommended.

According to the authors, the adopted solution is similar to the functioning of masonry units in the exterior surface finish. The obtained results confirm the usage of unprotected ceramic building materials designed for use in protected walls against water penetration.

Further studies are planned to analyze SEM, the distribution of soluble mineral salts and changes in the mechanical properties of the superficial ceramic layer. As a result of environmental influence, the processes of migration and crystallization of soluble mineral salts are also initiated, resulting in subsequent changes in the microstructure.

Author Contributions: Conceptualization, M.W. and A.K.; methodology M.W. and A.K.; investigation, M.W. and A.K.; data curation M.W. and A.K.; writing—original draft preparation M.W. and A.K.; writing—review and editing, M.W. and A.K.; visualization M.W. and A.K. All authors have read and agreed to the published version of the manuscript.

Funding: This research received no external funding.

Institutional Review Board Statement: No applicable.

Informed Consent Statement: No applicable.

Data Availability Statement: Data are contained within the article.

Conflicts of Interest: The authors declare no conflict of interest.

References

1. *EN 771+A1:201 2015-10;* Specification for Masonry Units—Part 1: Clay Masonry Units. British Standards Institution: London, UK, 2015.
2. *EN 1996-2:2006;* Eurocode 6—Design of Masonry Structures—Part 2: Design Considerations, Selection of Materials and Execution of Masonry. British Standards Institution: London, UK, 2006.
3. Koroth, S.R. Evaluation and Improvement of Frost Durability of Clay Bricks. Ph.D. Thesis, Concordia University, Montreal, QC, Canada, 1997.
4. Abu Bakar, B.H.; Wan Ibrahim, M.H.; Megat Johari, M.A. A review: Durability of fired clay brick masonry wall due to salt attack. *Int. J. Integr. Eng.* **2011**, *1*, 111–127.
5. Mensinga, P. Determining the Critical Degree of Saturation of Brick Using Frost Dilatometry. Master's Thesis, University of Waterloo, Waterloo, ON, Canada, 2009.
6. Hansen, W.; Kung, J.H. Pore structure and frost durability of clay bricks. *Mater. Struct.* **1988**, *21*, 443–447. [CrossRef]
7. Litvan, G. *Freeze-Thaw Durability of Porous Building Materials, Durability of Building Materials and Components ASTM STP 691;* American Society for Testing and Materials: West Conshohocken, PA, USA, 1980; pp. 455–463.
8. Ravaglioli, A. Evaluation of frost resistance of pressed ceramic products based on the dimensional distribution pores. *Trans. Br. Ceram. Soc.* **1976**, *76*, 92–95.
9. Robinson, G.C. The relation between pore structure and durability of bricks. *Ceram. Bull.* **1984**, *63*, 295–300.
10. Herget, F.A.; Crooks, R.W.; Winslow, D.N. Variability within single projects of physical properties of face brick as related to potential durability. In Proceedings of the 6th Canadian Masonry Conference, Department of Civil Engineering, University of Saskatchewan, Saskatoon, SK, Canada, 15–17 June 1992; pp. 417–428. Available online: https://www.worldcat.org/title/proceedings-6th-canadian-masonry-symposium-15-17-june-1992-department-of-civil-engineering-university-of-saskatchewan-saskatoon-saskatchewan-canada/oclc/70476722 (accessed on 18 March 2019).
11. Nieminen, P.; Romu, P. Porosity and frost resistance of clay bricks. In *Brick and Block Masonry;* de Courcy, J.W., Ed.; Elsevier Applied Science: London, UK, 1988; Volume 1, pp. 103–109.
12. Winslow, D. Predicting the durability of paving bricks. *J. Test. Eval.* **1991**, *19*, 29–33. [CrossRef]
13. Mallidi, S.R. Application of mercury intrusion porosimetry on clay bricks to assess freeze-thaw durability. A bibliography with abstracts. *Constr. Build. Mater.* **1996**, *10*, 461–465. [CrossRef]
14. Maage, M. Frost resistance and pore size distribution in bricks. *Mater. Struct.* **1984**, *17*, 345–350. [CrossRef]
15. Bellanger, M.; Homand, F.; Remy, J.M. Water behaviour in limestones as a function of pore structure: Application to frost resistance of some lorraine limestones. *Eng. Geol.* **1993**, *36*, 99–108. [CrossRef]
16. Scherer, G.W. Crystallisation in pores. *Cem. Concr. Res.* **1999**, *29*, 1347–1358. [CrossRef]
17. Abdrakhimov, Z.; Abdrakhimova, S. Chemical-elemental, phase composition and porosity structure of ceramic samples from Cham tower (Vietnam) more than 1000 y/o. *Glass Ceram.* **2018**, *75*, 33–38. [CrossRef]
18. Stryszewska, T.; Kanka, S. Forms of damage of bricks subjected to cyclic freezing and thawing in actual conditions. *Materials* **2019**, *12*, 1165. [CrossRef] [PubMed]
19. Elert, K.; Culturone, G.; Rodriguez-Navarro, C.; Pardo, E.S. Durability of bricks used in the conservation of historic buildings—Influence of composition and microstructure. *J. Cult. Herit.* **2003**, *4*, 91–99. [CrossRef]
20. Cultrone, G.; Sebastian, E.; Elert, K.; De la Torre, M.J.; Cazalla, O.; Rodriguez-Navarro, C. Influence of mineralogy and firing temperature on the porosity of bricks. *J. Eur. Ceram. Soc.* **2004**, *24*, 547–564. [CrossRef]
21. Tang, Y.; Shao, Z.; Xu, T. Pore structure of Ancient Chinese bricks under environmental vicissitudes. *J. Civ. Eng.* **2016**, *20*, 1895–1902. [CrossRef]
22. Kung, J.H. Frost durability of Canadian clay bricks. In Proceedings of the 7th International Brick Masonry Conference, Melbourne, Australia, 17–20 February 1985; Brick Development Research Institute, University of Melbourne, Department of Architecture and Building: Melbourne, Australia, 1985; pp. 245–251.
23. Koniorczyk, M.; Gawin, D.; Schrefler, B. Multiphysics for spalling prediction of brick due to in pore salt. *Comput. Struct.* **2018**, *196*, 233–245. [CrossRef]
24. *EN 772-11:2011;* Methods of Test for Masonry Units—Part 11: Determination of Water Absorption of Aggregate Concrete, Autoclaved Aerated Concrete, Manufactured Stone and Natural Stone Masonry Units due to Capillary Action and the Initial Rate of Water Absorption of Clay Masonry Units. British Standards Institution: London, UK, 2011.

MDPI AG
Grosspeteranlage 5
4052 Basel
Switzerland
Tel.: +41 61 683 77 34

Materials Editorial Office
E-mail: materials@mdpi.com
www.mdpi.com/journal/materials

www.ingramcontent.com/pod-product-compliance
Lightning Source LLC
Chambersburg PA
CBHW061621210926
51850CB00009B/2398
* 9 7 8 3 7 2 5 8 1 6 3 4 7 *